Free Will

A defence against neurophysiological determinism

John Thorp
University of Ottawa

ROUTLEDGE & KEGAN PAUL

London, Boston and Henley

First published in 1980
by Routledge & Kegan Paul Ltd
39 Store Street, London WC1E 7DD,
9 Park Street, Boston, Mass. 02108, USA and
Broadway House, Newtown Road,
Henley-on-Thames, Oxon RG9 1EN
Set in IBM Press Roman 10 on 12 pt by
Hope Services
Abingdon, Oxon
and printed in Great Britain by
Page Bros (Norwich) Ltd
Norwich and London
© John Thorp 1980

British Library Cataloguing in Publication Data

Thorp, John
Free will.
1. Free will and determination
2. Neurophysiology
I. Title
123   *BJ1461*   *80-40199*

ISBN 0 7100 0565 2

For my parents,
John Sidney Thorp
Anna Elizabeth Gannon Thorp

CONTENTS

Contents

FIGURES

ACKNOWLEDGMENTS

Something like this book was presented as a doctoral thesis in Oxford in 1976. I owe a very real debt of thanks to Rom Harré who was my patient and indulgent supervisor during most of its preparation, and to Sir Peter Strawson who undertook that office for part of the time; also to my gracious examiners John Lucas and Michael Ayers; and to John Ackrill who had faith in me and in this project in the early stages, when it was most needed. Anthony Kenny's egregious lectures on Will, Freedom and Power made me take compatibilism seriously – I had always thought it utterly beneath contempt – and the book is much, much better for that.

A number of people have said things in philosophical conversation – sometimes quite idle philosophical conversation – which have chanced to jolt my thinking on these subjects in important ways; some of them, I fancy, will be surprised to find themselves mentioned here: Arthur Adams, John King-Farlow, Bill Shea, Ralph Walker, Kai Nielsen, Jeremy Stone, and Pierre Laberge. For more focussed and more frequent discussions I am grateful to David Owen. My weekly neurological lunches in 1975-6 with Marc Colonnier were of enormous profit, and I miss them sorely now that he has gone to Quebec. The impetus for the discussion in chapter III of mental events and mental predicates came from the work of Naomi Scheman, though my views here are nothing like so carefully developed as hers; I am grateful for this impetus. Some of the material in chapter VI was read to the Société de philosophie de l'outaouais; I am grateful to that society for its invitation to speak, and to François Duchesneau and Phillip Rody for their penetrating comments on that occasion. William Dray and

Acknowledgments

Adam Morton read the whole manuscript with care at one stage, and provided me with very useful suggestions.

The copyright for my Figure 4.1 is owned by Springer-Verlag; I gratefully acknowledge their permission to reproduce that figure here.

I must also thank Cheslyn Jones for his generous hospitality and merry company on all my subsequent visits to Oxford; and Gilbert and Stewart Bagnani in whose serene house, 'Vogrie', a large part of the writing was done.

Finally, I must thank the Canada Council for supporting me when it did, and my parents for supporting me when it didn't. My debt to my parents during the time of my doctoral work goes far beyond this, though, and it is fitting that the dedication should be theirs.

I

INTRODUCTION

The problem of freedom and determinism is one of the most enduring, and one of the best, problems of philosophy; one of the best because it so tenaciously resists solution while yet always seeming urgent, and one of the most enduring because it has been able to present itself in different ways to suit the pre-occupations of different ages. At one time the problem arose because the operation of Fate seemed undeniable, at another because Divine omnipotence and human wrongdoing were the subject of great interest, at another because Divine foreknowledge seemed irrefragable, at yet another because psychological determinism was generally assumed to be true, and both first[1] and lately, it arises because a deterministic physical atomism has been the prevailing scientific theory. My concern here is with the problem in this oldest and newest of its versions.

In this version the problem first presents itself as a dilemma: either we are free and therefore nature is not deterministic, or nature is deterministic and therefore we are not free. The consequent of each of these horns imports embarrassment; for on the one hand we are, for certain reasons, inclined to believe that nature is deterministic, and on the other hand we do suppose ourselves free, and some of our moral and legal institutions seem to depend upon that supposition. Neither of these beliefs is, at first glance, impossible to give up, however: freedom *could* be an illusion, and it seems impossible to show either *a priori* or *a posteriori* that nature is deterministic. However that may be, there are always three ways of dealing with a dilemma: to embrace one horn, to embrace the other, or to attempt to reconcile the two. To reconcile the two horns of this dilemma we should have to show that the kind of

1

freedom we and our moral institutions require is not incompatible with natural determinism.

Most recently those who wish to assert natural determinism and pay the cost of denying freedom have not been vociferous; the field has been left to the other two: those who seek to work a reconciliation, and those who, thinking such a reconciliation cannot be effected, would wish to assert freedom and pay the cost of denying natural determinism. These are compatibilists and libertarians respectively. What they disagree about in the first instance is whether or not natural indeterminism is a necessary condition of freedom; the libertarian holds that it is, and the compatibilist that it is not – or even that it is a necessary condition of freedom that there not be natural indeterminism. I am inclined to think that on this particular issue it is the libertarians whose position is the more attractive, though the issue is not settled. However, there is more to it than this. Libertarians are typically short-sighted, and compatibilists are long-sighted. What looms in the distance for this metaphor is the question, what are the *sufficient* conditions of freedom? Unfortunately, libertarians always seem to go to pieces over this one; having insisted that a necessary condition of freedom is natural indeterminism it seems unconscionably difficult for them to say what the sufficient conditions may be; freedom is, after all, something more than mere randomness; when libertarians try to say what this something more is, their talk becomes at best evasive and obscure, and at worst incoherent. Compatibilists, being long-sighted, foresee this problem for libertarians and are therefore content to patch together a much less attractive position over the first question, in order to have an easier time over the second. Indeed, the best argument for compatibilism is still that libertarianism soon falls into obscurity and incoherence.

Libertarians are, however, typically dogged characters, and they cannot believe that they are wrong in their answer to the first question, namely, that natural indeterminism is a necessary condition of freedom; they prefer to hold on to that one little piece of, to them, self-evident truth, and to trust to time for the rest. What they require is an account of the sufficient conditions of freedom, given that one of the necessary conditions is natural indeterminism. It is just such an account that I attempt to provide in this essay: in a phrase, a libertarian theory of free decision.

What follows is, then, a defence of libertarianism. But it is a defence only in the rather conservative sense that it sets forth a coherent and unevasive version of that position; however, such is the sort of defence the position most needs.

If defence of a more offensive kind were requested, I should not know what to say. What could be the reasons for embracing a coherent and unevasive libertarian theory of free decision instead of a coherent and unevasive compatibilist one? The unattractiveness of the compatibilist view that what we mean by a free decision can none the less be a decision which is naturally determined would be my reason for choosing libertarianism, but I am well aware that to some that view is not unattractive: not all compatibilists are compatibilists only for want of a coherent statement of libertarianism. Perhaps the fact that libertarianism offers *more* freedom - freedom from more constraints - than does compatibilism would be something to be weighed in the balance; but it is not clear into which pan it should be put. One point that might be thought to weigh against libertarianism is its requirement of indeterminism in nature; but that objection is surely less weighty now than it must once have seemed: the postulation of (non-epistemic) indeterminism in physics is no longer anathema or even odd. Perhaps the choice would have to be made in the way that scientists choose between rival theories, honouring simplicity and conceptual economy, not only in the theories themselves, but also in the way in which they interlock with the 'prevailing scientific paradigms'; history does not, however, support the view that prevalence, in metaphysical paradigms, is a virtue. In short, I am stymied by this further question and shall confine myself in what follows to the nearer one: can a coherent libertarian theory of freedom be given?

1 VARIETIES OF FREEDOM

(i) Two capital distinctions

'Freedom', of course, can mean many things, and before the account can proceed with sanity it will be necessary to get a pretty clear idea of just what sort of freedom it is that I want to give a libertarian theory of. Some order can be made in the crowd of meanings and uses of 'free' if we notice that there are (at least) two capital distinctions which serve to divide them from one another. The first is the distinction among the items *to* which freedom may be attributed: act, will, choice, decision, person, etc. The second is the distinction which is tradition-

3

ally known as that between liberty of spontaneity and liberty of indifference. A cross-division of these distinctions will yield precisely the sort of freedom I shall be concerned with: liberty of indifference as it belongs to decisions.

About the first distinction I shall say nothing elaborative here, but only defend my interest in freedom of *decision*.[2] The source of philosophical worry about freedom is usually the troubling issue of responsibility. In debates about free will philosophers are in general hoping to make some room in the world for whatever kind and degree of freedom it is that our practice of holding people responsible for their acts requires. They disagree about just what kind of freedom *is* required for this purpose, but they do in general agree about the purpose. Now, people are usually held responsible for their actions rather than for their decisions; it may therefore seem odd of me to be concerned not with freedom of action but with freedom of decision. It would be less odd if all free responsible acts were preceded by free decisions to do those acts; but even the ghost of this old idea has been effectively exorcised now: many of the acts which I should want to call free and responsible are not preceded by any recognizable mental event having the character of a decision.

Still, some of them are. And for better or worse it is these acts, which are performed as a result of a full-scale careful decision to perform them, which have been regarded as paradigms of free action. Some libertarians, C. A. Campbell for example, hold that only such acts are free;[3] other libertarians, like Sartre, hold that all our acts are free. I incline to the latter view and will argue briefly for it in the final chapter. But in the meantime there is a considerable philosophical advantage in starting with full-blown decision-preceded acts. It is this: a condition of freedom in such an act is that the decision from which it results itself be free. Thus the libertarian, in order to give an account of the freedom of such an act, must give an account of freedom of decision. Now the really deep and difficult problem facing the libertarian faces him as well over free decisions as over free acts, but it faces him more squarely, with less ancillary clutter, in the case of decisions than in the case of acts. That is, the statement of sufficient conditions for a free action is likely to be more complicated and longer and more disjunctive than the statement of the sufficient conditions for a free decision – simply because there is so much room for things to go wrong between a decision and the action which embodies it. The philosophical advantage of being concerned with decisions rather than actions is that,

4

for the moment, one can forget about those things that may go wrong. Restricting the enquiry to decisions has, if you like, the advantage of isolating the main problem. It is for this reason that what I seek to give is a libertarian account of free *decision*; in my last chapter I shall suggest how the account is to be extended to cover actions as well.[4]

The second capital distinction is that between liberty of indifference and liberty of spontaneity. The distinction is freshman stuff, but I want to explore it a little. I do this not in order to go through the philosophical ritual of making obvious distinctions, but because it often happens that libertarians, when they get to the really difficult moment in making their case, covertly give up the pursuit of liberty of indifference, and settle for a version of liberty of spontaneity. This is a shift which must be guarded against.

Liberty of spontaneity belongs to a person when he is not somehow hindered from expressing his wants in action. By extension it can be predicated of a person's act, when that act expresses the agent's wants, or his rational self, or his real nature, or some other psychological item. It is thus a kind of liberty which is not essentially bound up with choosing or deciding, with the presence of alternatives. But by further extension it *can* be predicated of a person's choice or decision when the outcome of that choice or decision expresses the agent's wants or his rational self or his real nature, etc. (It is in this last extension that defeated libertarians invoke it.[5])

Liberty of indifference, on the other hand, belongs to a person when he is able to make a choice, when there are genuine possibilities before him. Retrospectively, it can be predicated of a person's act when an affirmative answer can be made to the question, 'could he have done otherwise?'; and it can be predicated of a person's decision or choice when an affirmative answer can be made to the question, 'could he have decided/chosen otherwise?'

In the case of decision, the distinction between the two kinds of liberty can be adumbrated in another way. The English word 'decision' has two meanings, stemming from two different etymological sources: *decisio,* the act of deciding, and *decisum,* the outcome of the *decisio,* that which is decided upon. In the same way 'choice' can mean either the act of choosing or the alternative chosen. Liberty of indifference is predicated retrospectively in virtue of a property of the *decisio,* namely, that there really were alternatives open; and liberty of spontaneity is predicated in virtue of a relational property of the *decisum,* namely, that it accords with the agent's desires, his rational self, his real nature, or whatever is of interest at the time. This explains how it can be that we meet occasionally with anomalous reflections like the following: 'He is free now either to go on in his merry old way or else to make a clean break and reform his character; of course the latter would be the really free thing to do.' This sort of remark seems to imply that only if one chooses one of the things one is free to choose is the choice really free; if one chooses the other thing one is free to choose, the choice is not free. And this is very odd. The first sort of freedom is liberty of indifference and it has to do with the *decisio,* and is not affected by what the *decisum* is; the second sort is liberty of spontaneity and it has to do with the *decisum*: it can perfectly well belong to one possible *decisum* but not to others.

It is also to be noticed that liberty of spontaneity is a highly relative affair. On any occasion of its attribution one must have in mind (a) what constraints they are whose absence is being signalled,[6] and (b) what it is in the agent that is thus being allowed free expression. One can exhibit just how relative this sort of freedom is by giving two cases in which the referents of (a) and (b) are exactly reversed. If I drink too much at a party, lose all rational prudence, and make passionate declarations of love, I might say that I was free; here the constraints of reason are absent and it is desire that gets expressed. Yeats and D. H. Lawrence and Kazantzakis would exalt this case of freedom. In another case I might find that my desire has waned and that I am able to conduct my amorous relationships with people more rationally, more calmly. This, too, I might describe as a case of freedom: but the absent constraint here is desire, and what gets expressed is rational wish. Countless persons, from Cato to the Archbishop of Canterbury, would exalt this case of freedom. It is a truism that our rational and our passionate natures are often in competition. When either is in the ascendant we may come to regard ourselves as free, free from subjection to the other.

The unhindered expression of many different things may, on different occasions, count as liberty of spontaneity: of surface desires, of the moral self, of the rational self, of deep longings, of the agent's 'true nature', of subconscious desires, of past character, etc. It may even be the expression of items not straightforwardly within the agent at all that counts as freedom; this perhaps explains the otherwise baffling religious adage that perfect freedom consists in complete submission to the will of God. This extreme relativism of liberty of spontaneity makes it a slippery philosophical chess-piece. I do not deny that some kinds of liberty of spontaneity are a necessary condition of an act's being considered free and the agent responsible for it; but it does not seem, prima facie, as though any sort of liberty of spontaneity will belong in the account of what is necessary for a *decision* to be free, in the sense that holding the agent responsible for it requires. This can be shown conclusively by the following consideration. An agent may freely and responsibly decide *against* the promptings of his surface desires, of his moral self, of his rational self, of his deep longings, of his 'true nature', of his subconscious desires. It is true that in a minimal sense liberty of spontaneity is a necessary condition of freedom in the sense required by responsibility: a free decision must express *something* in the agent, whether it be deep desire or momentary whim; but it does not seem that there is enough strength in this condition to warrant its exploration. The proof of its weakness is that it is also a condition of unfree decisions: a person making decisions whose outcome has been fore-ordained by hypnosis nonetheless expresses something in himself. It seems as though this minimal kind of liberty of spontaneity is a necessary condition of free decisions only because it is a necessary condition of decisions.

Liberty of spontaneity is not, then, a necessary condition for a decision to be free in the sense that responsibility requires. Except in a vague and unhelpful way, it does not seem to be bound up with freedom of decision at all. The sort of liberty we have in mind when we speak of decisions seems rather to be liberty of indifference. A decision is free if the agent could have decided otherwise; and if we are satisfied that he could have decided otherwise we are by and large prepared to hold him responsible for his decision.

For these reasons, then, my concern is with liberty of indifference as it belongs to decisions: what conditions are necessary for the truth of 'he can decide either way' or of 'he could have decided otherwise'? Hereafter, when I speak of free decision it will be liberty of indifference that I have in mind.

(ii) Scalar and absolute freedom

There is another distinction which should be brought in to organize our ways of talking and thinking about freedom: sometimes we speak as though there can be degrees of freedom and sometimes we speak as though there cannot. I shall not try here to give a general analysis of these two ways of thinking, but only enquire whether and how they operate in the case of free decision. At first it might seem that scalar freedom has no place here: the test for freedom in a decision is 'could he have decided otherwise?', and, whatever that question means, it seems to be one which admits only yes or no as an answer. And yet on the other hand someone could quite naturally plead that while, yes, one *could* have decided otherwise than one did, any other decision than that which one took would have been so difficult and so wearing to implement, that really one should not be thought of as having been wholly free in deciding as one did.

How are these two intuitions, on the one hand that freedom of decision is absolute, and on the other that it is scalar, to be reconciled? My way of doing this is to offer a fairly drastic theory; the compensation for its drastic-ness is a tidy organization of these matters.

The theory is that although we do very often speak as though strain, duress, difficulty, extreme temptation, torture, and influence in general *lessen* freedom, this is only a *façon de parler*. Freedom of decision is properly to be regarded as absolute, an all-or-none affair. The *façon de parler* arises from the fact that we sometimes think of responsibility as scalar, and since freedom and responsibility are much intertwined concepts, we are inclined to suppose that a case of diminished responsibility is a case of diminished freedom. But I should further argue that there is something very odd also about the idea of scalar *responsibility*: responsibility is properly absolute, just as freedom is – either one is responsible or one is not. So that 'diminished responsibility' is itself a *façon de parler*. How then does this particular *façon de parler* arise? Well, when we say that a man's responsibility for a heinous crime is diminished, that is to be read as that a man is responsible for a crime of diminished heinousness. On this theory, then, the difference between a case where a man deliberately, coolly and calmly commits murder and a case where he commits murder in a frenzied passion of jealousy or anger is *not* that the second man's responsibility for murder is diminished; both are equally responsible for their crimes, but the crime of the second is a less heinous one. The crime is not properly describable as 'murder' in both cases; in the first case the crime is 'cool, deliberate murder', and in the second the crime is properly described as 'murder committed in a passion'. The second is a less heinous thing than the first. An alleged case of diminished responsibility is really, on this theory, a case of responsibility for a crime of diminished moral enormity. Now, it is the presence of just such items as strain, duress, difficulty, extreme temptation or torture which will diminish the moral enormity of a crime in this way. Thus it arises that we come to think of the presence of these items as diminishing first responsibility and secondly freedom.

This theory, which I have indeed very briefly sketched, will be elaborated in chapter VII. I proceed, then, on the assumption that freedom of indifference is an absolute, all-or-none affair.

(iii) Complete and incomplete freedom

There is one further complexity in the conceptual terrain surrounding the idea of freedom, about which it is important to be clear. It often seems to us that we have choices to make, and these choices are usually among a limited number of alternatives. Freedom to decide among a limited number of alternatives I shall call incomplete freedom, without meaning to suggest that it is in any way an unsatisfactory or inadequate sort of freedom. By contrast, complete freedom would be the freedom to decide or do anything at all. It seems to be part of Sartre's extreme libertarian view that freedom is complete in this sense. I do not follow Sartre in this; it is no part of my intention to make a conceptual place for the completeness of our freedom; I do not believe that at any one moment in time absolutely any decision is possible for a person. Logic, physics, history, and unimaginativeness all limit the range of alternatives open to us at a given moment: our freedom is in this way incomplete. What I wish to argue is that among those limited alternatives we have freedom of choice. I am concerned, then, to establish the conceptual possibility of incomplete but absolute (all-or-none) freedom of indifference belonging to decisions.

The point about incompleteness is important in this way. One can be put into a mood of deterministic wistfulness by the reflection that whatever freedom we have to navigate among our surroundings and our desires, proclivities, and talents, the general tenor of our lives is set by those surroundings, desires, proclivities, and talents. Russell's powerful remark that however free we may be to do as we please we are not free to please as we please makes just that point. We inherit our desires and our preoccupations, and these, among other things, limit the range of alternatives before us at any time. We are thus victims, one might say, of incomplete determinism at least. What I am concerned to argue is that we are not victims of complete determinism: the range of alternatives before us may be determined, but which one among the range we choose is not. The mood of wistfulness that is provoked by ruminating on the truth of incomplete determinism is not a mood that my argument seeks to dissolve.

2 VARIETIES OF DETERMINISM

Determinism is the doctrine that what happens must happen – that all true propositions recording events in the universe are necessary pro-positions and all false ones impossible. There have been various ways of justifying this universal insertion of the modal operator for necessity, and these have constituted the various sorts of determinism: fatalism, theological determinism, psychological determinism, and physiological determinism, to name the usual four.

Fatalism and theological determinism will not be my concern here. They are engaging and instructive problems, but the philosophical fashions which prevail make them seem not urgent. The God-hypothesis seems not now very widely accepted in an uncompromising form; and while the views about time which make fatalism work *are* widely accepted, they are accepted without being firmly believed: we are quite prepared to be told that our commonsense views about the nature of time, or about tensed logic, need to be revised.

Similarly, psychological determinism does not any longer seem press-ing, on its own merits. For one thing it seems to be false in fact: no one succeeds or even nearly succeeds in producing plausible psycho-logical laws – laws, that is, that are fine-grained enough to capture all individual decisions and intentions in their net. Furthermore, there seems no clear reason why there should be any analogy between the way series of contiguous mental events are related and that in which series of contiguous physical events are related; psychological determin-ism has usually rested on the supposition of such an analogy. On the other hand it has been argued that if neurophysiological determinism is true then psychological determinism must be true, not by analogy but by deductive inference. We shall look at this claim shortly.

Physiological determinism in its various forms is the most urgent and plausible of the four. We must distinguish (i) physiological determinism, the thesis that every bodily event is uniquely determined by physical causes; (ii) determinism in neurophysiology, the view that every neuro-physiological event is uniquely determined by physical causes; (iii) neurophysiological determinism, the view that every mental event is uniquely determined by physical causes. View (iii) is entailed by (ii) together with the Correlation thesis, that there is a lawlike correlation between mental event kinds and neurophysiological event kinds. It is (iii), neurophysiological determinism, that will be my chief concern, for it is the one which is at odds with freedom of decision.

We can usefully begin the study of neurophysiological determinism by examining an initial objection to its truth. Here is a train of argument that has sometimes been used by strong disbelievers in psychological determinism: (1) if neurophysiological determinism is true, then psychological determinism is true; (2) but psychological determinism is obviously false; therefore (3) neurophysiological determinism is false; therefore (4) either determinism in neurophysiology or the Correlation thesis is false; but (5) determinism in neurophysiology is true; therefore (6) the Correlation thesis is false. My objection to this argument is an objection to the first premise.

One can easily see the line of thought that leads one to suppose that premise true. If each mental event kind is correlated with one neural event kind, then the lawlike chain of neural events will be perfectly matched by a similarly lawlike chain of mental events. This is illustrated in Figure 1.1.

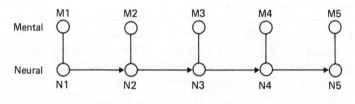

Mental   M1   M2   M3   M4   M5

Neural   N1   N2   N3   N4   N5

t ⟶

Figure 1.1 Neurophysiological determinism: first approximation

Just as N5 is made inevitable by the occurrence of N4, so the occurrence of M5 is made inevitable by the occurrence of M4. The chain M1-M5 will be as lawlike as the chain N1-N5.[7]

But this is to misrepresent the thesis of determinism in neurophysiology. That thesis does not claim that every neurophysiological event is uniquely determined by preceding neurophysiological events; rather it claims that every neurophysiological event is uniquely determined by preceding *physical* events. The former would be true if neurophysiology were a closed system, but of course it is not a closed system. Many other sorts of physical events bring about changes in the neural state. The case would be better represented as Figure 1.2.

11

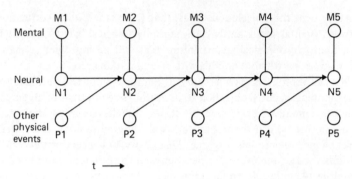

Figure 1.2 Neurophysiological determinism: second approximation

Just as, according to this scheme, N1 is not a sufficient condition for the occurrence of N2, so M1, which is correlated necessarily with N1, is not a sufficient condition for the occurrence of M2. M1 will yield M2 only if certain other physical events (P1) occur. Since the thesis of determinism in neurophysiology does not assert that whenever N1, then N2 immediately afterwards, it cannot entail that whenever M1, then M2 immediately afterwards.

An example may help to make this clear. Suppose that M1 is the state of having a headache, and M2 a moment in the process of the disappearance of the headache – the sense of its lessening. N1 will be the neural correlate of having a headache, and N2 the correlate of losing it. At t_1 the subject ingests aspirin (see Figure 1.3).

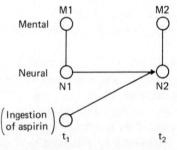

Figure 1.3 An example of the second approximation

N1, together with the ingestion of aspirin, brings about N2. And given that N2 occurs, M2 must also occur (if, that is, the Correlation thesis is true). But there is no necessary correlation between N1 and the

12

ingestion of aspirin: so the occurrence of N1 does not lead inevitably to the occurrence of N2. Hence the occurrence of M1 does not lead inevitably to the occurrence of M2.[8]

Thus the fact that psychological determinism is patently false does not at all show that neurophysiological determinism is false.

There are several crude variants of the thesis of neurophysiological determinism which are falsifiable by experience or experiment. It is worth spending a paragraph to make it clear that the one which concerns me is not one of these. There are variants, for example, which suppose that the thesis entails that usually our bodies move in certain ways without our wanting them to, or against our wills. Such, for example, was the determinism of the ancient Stoics. More modern versions have thought that the thesis entails the impossibility of learning or of adapting: these might be called the single-level deterministic theories. But clearly the neurophysiological apparatus is a highly sophisticated homeostatic or homeorhetic apparatus which is capable of producing adaptive behaviour, even if the neural mechanism of adaptation is still imperfectly understood. What is more, such a homeostatic neural apparatus may perfectly well be deterministic, as all other homeostatic apparatuses are. So that the deterministic thesis we are entertaining is not one which is falsified by the perfectly correct observation that men sometimes change their ways or their characters. What the thesis will say to that observation is just that the decision to make the change had, of course, a neural correlate, and that correlate was made to occur by preceding physical, including neurophysiological, events. So every mental event is uniquely determined by physical causes.

Indeed it seems important to say that the thesis of neurophysiological determinism does not fall foul of any experience whatsoever. It is entirely capable of saving all the phenomena, where the 'phenomena' are understood as observables. In this way it is a highly sophisticated theory, and it may very well be true. What it *does* fall foul of is a certain portion of the metaphysics that is built into our ordinary ways of thinking about decision, action, and responsibility; even this claim is contentious, though, and it will be defended in chapter II. Freedom is not at all an obvious fact: the feeling of freedom could very easily be illusory.

There is another line of argument against neurophysiological determinism whose propositions may well be true, but which falls wide of its target. It is said, for example, that however proficient we may

13

become at explaining men's actions with a neurophysiological story, that sort of explanation will never supplant, never do the job of, ordinary sorts of explanation in human terms. It is also said that the ordinary sort of explanation is much more interesting than the neurophysiological sort would be: it tells us more of the kind of thing we want to know. It is also argued that teleological explanations of human action are not reducible to neurophysiological ones, or that the correctness of a neurophysiological account of a man's action would not preclude its being true that he did what he did for certain reasons. All of these propositions may well be true. But neither alone nor together would they show that neurophysiological determinism is false, nor, as I think, would they remove the difficulties which it entails.

For it is no part of the thesis of neurophysiological determinism that a neurophysiological explanation of a human action would render any other sort of explanation of it either false or uninteresting. Thus it may perfectly well be true, for example, both that a man decided at t to ϕ because he had a sudden urge to ϕ, and *also* that physical events around and in his nervous system prior to t made it inevitable both that he had a sudden urge to ϕ and that he decided at t to ϕ. There is no more difficulty about this than about saying that a Türing machine changed the subject because when it couldn't answer my question it tried to hide its ignorance, and also that it changed the subject because the state of its mechanism together with the most recent inputs to it, made it inevitable that it should.

But although neurophysiological explanations would be perfectly compatible with teleological explanations, they would in a sense override them. Neurophysiological explanations would be *stronger* than the other sort, and that in two ways: they would render the explanandum as unique, and they would render it as necessary.

Teleological explanations do not typically render the explanandum as unique. By that I mean that a given teleological explanation would be apt for quite a range of different possible explananda. Suppose that an agent opens a door; we ask why, and are told that he wanted to go into the room – a teleological explanation. But *that* explanation does not tell us why precisely he opened the door; for it would also do as an explanation of opening the French window, the trap door, or the skylight. Not so with the sort of neurophysiological explanation we are envisioning: there the causal story together with the Correlation thesis explain why precisely that act was done which was done: the explanation will not do for any other possible act or decision. Two objections will be made to this contention: first, that *sometimes* teleological explanations do render their explananda as unique, and second, that, anyway, their usual failure to do so arises only because we do not typically give them in a fully articulated form: if we did articulate them fully they would render the explanandum uniquely. And it is unfair to compare the power of actual teleological explanations, offered in the hustle and bustle of life, with that of *idealized* neurophysiological explanations; rather we must compare the idealized versions of both.

The first objection has in mind the sort of case where we say that a man took the 2.15 train to London because he had to be there by 4.00 and could not leave earlier than 2.00: it was the only possible train for him to take. But of course even this does not render the action as unique, for he could have driven there, been driven there, taken a fast coach, a helicopter, or a balloon. It won't often happen that all *these* alternatives are excluded in a teleological explanation of taking a certain train. So that very often explanations which seem at first to render the explanandum uniquely will be seen, on closer inspection, not in fact to do so. But surely there are *some* which do. Suppose a man goes up the Eiffel Tower and we offer this teleological explanation of his action: he wanted to go up the Eiffel Tower. Clearly, there are no competitors with going up the Eiffel Tower for an action which satisfies a want to go up the Eiffel Tower.

Here, then, the teleological explanation does render the action in question as unique. But here the neurophysiological deterministic explanation would override the teleological in another way: it would render that action as unique under all the descriptions under which it is intentional. That is, 'he wanted to go up the Eiffel Tower' does not explain why he went up by the stairs instead of by elevator – but a neurophysiological explanation would explain that. However, the action under the description 'going up the Eiffel Tower' *is* rendered uniquely by the teleological explanation 'he wanted to go up the Eiffel Tower'. This admission, of course, does not cripple my contention, for I claimed only that teleological explanation does not *typically* render its explanandum as unique.

The second objection was that in their idealized version teleological explanations *do* render their explananda as unique. What this comes to, I suppose, is the claim that for any action a teleological explanation which renders that action as unique can be found. Here I think that the objector stands a much better chance of making his case. In the example of going to London by train we can rule out all other means of getting there by saying 'he wanted to travel in what for him was the cheapest way'; in the case of going into the room one can rule out all other means than the door by saying 'he wanted to enter in the easiest way'. Doubtless there are some, even many, cases of actions which finally contain an element of whimsy not amenable to teleological capture, but the champion of teleological explanation can, I think, with justice say that many actions can be rendered as unique by ideally complete teleological explanations.

However, here the second and greater strength of deterministic neurophysiological explanations comes into play. Teleological explanations *never* render their explananda as necessary; deterministic neurophysiological ones always do. A deterministic neurophysiological explanation renders its explanandum as naturally necessary, as something which under the laws of nature *had* to happen. Now a teleological explanation may sometimes show that its explanandum is a necessary condition for the attainment of a certain end; but since such an explanation, in so far as it is teleological, is never in a position to say that the end will be attained or must be attained, it cannot show that the explanandum was necessary. Thus if I say 'he turned the switch in order to light the lamp' I have given a teleological explanation; but if I say 'it was a necessary condition of his lighting the lamp that he turn the switch, and he lit the lamp, ergo necessarily he turned the switch' I have rendered his turning the switch as necessary (historical necessity?), but I have not given a teleological explanation.[9]

It is finally the natural necessity with which neurophysiological determinism would endow our decisions which makes those explanations stronger than the teleological ones which we readily accept, and which constitutes the threat of determinism. For we do not suppose that our decisions are naturally necessary occurrences; indeed we suppose that they are not. This will be argued in chapter II.

3 FORMAL MATTERS

Crucial to the argument of the previous part was the claim that a deterministic neurophysiological account of a man's decision would render the making of that decision as necessary. This claim, however, has still to be made good. There is a problem in its way, but the embarrassment it occasions is, I am inclined to think, an *embarras de richesses*: there seem to be several good lines of solution for it. We shall examine the problem and four possible solutions.

(i) The problem

The thesis of neurophysiological determinism is derived by syllogism from its two tributary theses, the thesis of determinism in neurophysiology and the Correlation thesis, in the following way. Where P is a certain kind of total physical state of and surrounding a man, N a certain kind of neurophysiological state, and D a certain kind of decision, t is any time, and t' is at a fixed interval later than t, and $\square$ denotes natural necessity, we have:

(1) $\square$ [(P at t) $\rightarrow$ (N at t')]   determinism in neurophysiology

(2) $\square$ [(N at t') $\rightarrow$ (D at t')] [10] Correlation thesis

∴(3) $\square$ [(P at t) $\rightarrow$ (D at t')]   syllogism, (1) and (2)

Then (3) is a law which the neurophysiological determinist accepts as true,[11] and which he will employ in giving his neurophysiological explanation of a man's decision. The ready inference by which, using this law, he will hope to demonstrate that an individual decision was necessary is the following one. Suppose that at t_1 a man takes a decision of the kind in question – say a decision to divorce his wife; the neurophysiological determinist notices that at t_0 the total physical state in and surrounding the man was of kind P; and he argues as follows:

From the general rule (3) he infers the particular application of it (3'):[12]

(3') $\square$ [(P at t_0) $\rightarrow$ (D at t_1)] [13]

(4) P at t_0

∴(5) $\square$ (D at t_1)

deducing thus that the man's decision at t_1 to divorce his wife had necessarily to occur.

Now the inference (3'), (4), ∴ (5) is of the form:

(i) $\square$ (p $\rightarrow$ q)

(ii) p

∴(iii) $\square$ q

a form which exhibits the well-known fallacy of deriving necessity of consequent from only necessity of consequence. The argument showing that the decision was necessary is, then, not valid.

Of course, this problem is not peculiar to the issue of neurophysio-

17

logical determinism. It arises for any alleged case of natural necessitation whatever. For example, one might believe that it is – or is entailed by – a law of nature that if in ordinary conditions I let go a heavy object one metre above the earth's surface (p), then it follows necessarily that it will drop to earth (q). That is, I might believe

(i) $\Box\,(p \rightarrow q)$

But suppose that in ordinary conditions I drop such an object, that is, suppose

(ii) p

I shall not be able to conclude

(iii) $\Box\,q$

that is, I shall not be able to conclude that it is necessary that the object drop to earth.

This result is of course surprising and counter-intuitive, for having let go such an object in such conditions, and believing (i) to be true, I would quite naturally say that it *must* fall to earth. Can our intuitions here be reconciled to the canons of modal logic? Unless they can be so reconciled no causal determinist thesis can show that any individual event had or has to happen, and all such determinist theses are disarmed at once.

(ii) The first solution: temporal modal logic

The first solution I wish to examine and propose would make the assertoric second premise (ii) necessary with the necessity of present and past fact[14] and would thus render the argument valid. To symbolize this proposal we must introduce dating notation as follows. Since the variables in the cases we are now considering denote tokens or types of events, we shall date them with t_0, t_1, t_2 etc. for tokens, and t, t′, t″ etc. for types of events. The modal operator for necessity will similarly be dated, with t with subscript when what falls within its scope is a sentence about events, and with t with superscript when what falls within its scope is a sentence about kinds of events. The expression $\Box t$ is to be read 'it is necessary at t', and $\Box t_+$ is to be read 'it is necessary at and after t'. We shall take it as axiomatic that:

A1 $pt \rightarrow \Box t_+\, pt$

or:

if p occurs at t, it is necessary at and after t that p occur at t,

and A2 $\Box\, p \rightarrow \Box t\, p$

or:

if p is necessary without reference to time, then p is necessary at any time, t.

The invalid argument (i), (ii), (iii) can now be reformulated as follows:

(iv)	$\square\,(pt \to qt')$	reformulation of (i)
$\therefore$ (v)	$\square\,(pt_0 \to qt_1)$	general rule applied to particular case
$\therefore$ (vi)	$\square t_{0+}\,(pt_0 \to qt_1)$	A2, (v)
(vii)	pt_0	reformulation of (ii)
$\therefore$ (viii)	$\square t_{0+}\,pt_0$	A1, (vii)
$\therefore$ (ix)	$\square t_{0+}\,pt_0 \to \square t_{0+}\,qt_1$	by standard modal rules
$\therefore$ (x)	$\square t_{0+}\,qt_1$	(viii), (ix), *modus ponens.*

Thus we conclude that it is necessary at and after t_0, the time at which the object is let go, that it drop to earth at t_1.

Of course, before adopting this solution to the problem, one would want to have a consistent and detailed dated modal system along these lines worked out. I shall not try to offer such a system here, but only say two things about it: one to notice a way in which such a system as this is better than the modalized tense logics which have been elaborated so far (at least, the ones I know of); and one to notice a difficulty which this system, as employed above, will have to face.

The first point is this. If one is going to solve the problem in hand by using modalized temporal logic and the principle of the necessity of past fact, then the calculus one uses will have to be one in which both the variables *and* the modal operators are given dates. The reason for this is as follows. Suppose that at t_6 in the past, a man took the decision to divorce his wife. It is not enough, in making the determinist claim, to say that it is necessary that at t_6 the man so decided – for that is true just by necessity of past fact, and will be allowed by the most ardent libertarian. What is needed, rather, is to be able to say, now, that it was necessary at some time before t_6 that at t_6 the man decide to divorce his wife. It is the claim that an act was necessary before it happened that sets the determinist debate going. And to make the latter claim the modal operators must be dated. To my knowledge no system with this feature has been elaborated.[15]

The second thing I wish to notice is this. If such a system of temporal modal logic were made and if it were used in arguments like (iv)-(x) above, one of the issues that would need discussion is that of the propriety of mixing modalities – or rather that of mixing kinds of modality – in one argument. For in the above argument the necessity of (iv), (v), (vi) and (ix) is natural necessity, and that of (viii) is necessity of past fact. In such an argument what sort of necessity attaches to the conclusion? One's inclination is to adapt Theophrastus' rule *peiorem semper sequitur conclusio partem* and say that the necessity belonging to the conclusion is of the weakest sort that occurs in the premises. And that of course is to ask for a comparative rating of the strengths of the different sorts of necessity (as well as the different sorts of possibility), and *that* is to ask for a big metaphysical job to be done.

(iii) The second solution: Barbara LXL

The second solution argues that the inference schema (i), (ii), ∴ (iii) is valid as it stands.

Controversially, Aristotle admitted as valid the mood Barbara LXL; that is, he allowed that one could derive a necessary conclusion from a necessary major premise together with a merely assertoric minor premise.[16] There has been a long history of debate whether Aristotle was right about this, but if he was right our problem is solved: I shall show that the inference schema (i), (ii), ∴ (iii) is equivalent to Barbara LXL.

Barbara LXL is of the form:
(a) Necessarily, all B's are A.
(b) All C's are B.
∴ (c) Necessarily, all C's are A.

We can read (i) □ (p → q) as 'necessarily, all spatio-temporal regions which contain p are spatio-temporal regions which contain q' or, more simply, 'necessarily, all p-regions are q-regions'. (ii) can be read 'x is a p-region', and this is formally parallel to (b) above if we allow the usual practice of classical logic in regard to singular propositions, namely, treating them as universal ones.[17] And similarly (iii) can be read 'necessarily, x is a q-region'. So:

(i′) necessarily, all p-regions are q-regions.
(ii′) x is a p-region.
∴ (iii′) necessarily, x is a q-region.

is valid by Barbara LXL.

But what about this mood? Is it to be considered valid? Many, certainly, have thought not, and the reasons are obvious. Łukasiewicz, however, well expressed the sense that it is valid in this ingenious remark:[18]

The following picture will perhaps make the syllogism . . .

[CLAbaCAcbLAca] more acceptable to intuition. Let us imagine that the expression LAba means: 'Every b is connected by a wire with an a'. Hence it is evident that also every c (since every c is a b) is connected by a wire with an a, i.e., LAca. For whatever is true in some way of every b is also true in the same way of every c, if every c is a b. The evidence of the last proposition is beyond doubt.

Now I cannot here go into the long history of the dispute about Barbara LXL; but the most recent substantial contribution to it, that of Rescher,[19] seems, if silence betokens consent, to have had wide acceptance. Rescher's view is that Barbara LXL is valid, but that the interpretation showing it to be so cannot be expressed formally, but only informally: the mood is valid because the major premise lays down a necessary rule of some sort and the minor describes some special case which falls under this rule. Storrs McCall has completed and confirmed this interpretation by showing that and how it applies to other moods with one apodeictic and one assertoric premise.[20] Conspicuously, the case we are considering is one in which the assertoric minor premise picks out a special case of the apodeictic general rule enunciated in the major. So that if Rescher's interpretation and defence of Barbara LXL is correct, our formal problem is solved.

It is worthy of note, however, that Rescher, giving an example of the Barbara LXL whose validity he has just defended, writes: 'If all elms are necessarily deciduous, and all the trees in my yard are elms, then all trees in my yard are necessarily deciduous (even if it is not necessary that the trees in my yard be elms).'[21] It looks as though what Rescher has here defended is *not* an inference of the form

(a) necessarily, all B's are A.
(b) all C's are B.
∴ (c) necessarily, all C's are A.

What has been defended seems rather to be

(a) necessarily, all B's are A.
(b) all C's are B.
∴ (c′) all C's are necessarylly-A.

changing the modality of the conclusion from *de dicto* to *de re*.

But once we have seen this, we must also see that Rescher's wording of his example would equally admit the modality of (a) as well as that of (c) to be *de re* rather than *de dicto*. But if that is what is intended, then there is no Barbara LXL in play at all, and the syllogism which has been defended is a straightforward Barbara XXX:

(a′) all B's are necessarily-A.

(b) all C's are B.

∴ (c′) all C's are necessarily-A.

We can however presume that this last is not the form of inference that Rescher was hoping to defend, for it is a form of inference that needs no defence. If, as seems more likely, it is (a), (b), ∴ (c′) that Rescher is wanting to make good, then a short way to unquestionable success would be to show that (a) → (a′), or in modern symbolism, to show that □ (p → q) → (p → □q). In section (v) below I shall argue that something like this principle might be allowed for the logic of natural necessity.

To recapitulate: if, as many have thought, Barbara LXL is a valid mood of the syllogism, then our problem about detachment in arguments involving natural necessity is solved. We are, however, beginning to suspect that there is a confusion between modality *de dicto* and *de re* involved in the claim that Barbara LXL is valid.

(iv) The third solution: relative modality

G. H. von Wright has developed a system of relative modality which he takes to be apt for symbolizing, among other things, arguments involving natural necessity.[22] Its peculiar formula is M(p/q), which is read 'p is possible on condition q' – and from this is defined N(p/q), 'p is necessary on condition q'. Absolute necessity is then defined in terms of relative necessity: 'p is absolutely necessary' = df. N(p/t), where t is any tautology.

This system has no provision for detachment of the conditional. I shall argue that this lack makes it unfit for symbolizing arguments about natural necessity that do in fact go on. I shall also suggest a way in which the system might be adapted to make good this lack.

The argument that this calculus is inadequate can take the form of an imagined conversation between two people, W, who never says anything that can't be symbolized in the von Wright system, and A, who speaks and reasons normally. Let them be malefactors who have just removed the main bolt from a bridge and are waiting to see it collapse.

A: If the bolt has been removed the bridge must collapse. And we have just removed the bolt. So the bridge must collapse.

W: No! The bridge must collapse provided the bolt has been removed.

A: Yes, but we *have* removed the bolt.

W: I agree, so, the bridge must collapse if the bolt has been removed.

A: But since we *have* removed the bolt it follows that the bridge must collapse.

W: As long as the bolt has been removed, it must collapse.

A: But the bolt *has* been removed.

W: I know.

A: Can't you see then that the bridge must collapse?

W: No, all I can see is that *if* the bolt has been removed the bridge must collapse.

This much will suffice to show how extremely odd and artificial it is to refuse detachment in the case of (at least) natural relative necessity. What are the reasons for this refusal? What is it that is being preserved at this high cost of artificiality? The device seems to be one that is designed not to let us forget that the necessity in question is *relative*, by never letting us suppress mention of that to which it is relative. There might be other ways to underline the relativeness of this necessity – ways which do less violence to our ordinary manner of reasoning. I shall be suggesting a fairly radical such way in the next section.

First, however, I wish to suggest a less radical one, one which will pretty much keep the prohibition on detachment, but which will none the less allow us to speak a little more normally. Traditional grammar distinguishes between conditional clauses on the one hand, and causal ones on the other. The former are introduced, in English, by such conjunctions as 'if', 'provided that', 'as long as', etc., and the latter by 'since', 'whereas', 'seeing that', 'inasmuch as', etc.[23] We can introduce a new dyadic operator ':', such that N(p:q) is to be read 'p is necessary, since q'. We might then permit the inference:

(1) N(p/q)

(2) q

∴ (3) N(p:q)[24]

W, in the above imaginary conversation, would have been much less intolerable had he been insisting that the bridge must collapse *since* the bolt had been removed: such a calculus as this will permit something like our ordinary reasonings about natural necessity. And this calculus will of course be strong enough to make the determinist claim: 'the decision to divorce his wife had to occur at t_6 since such and such physical state occurred at t_5'.

It might be thought, however, that the calculus is too strong, for it will admit even what von Wright calls Aristotelian 'self-necessity',[25] the principle that if p is true, then it is necessary that p be true: $p \rightarrow N(p/p)$.

That is, it can be shown of any proposition whatever that it is relatively necessary; it can be shown of any event whatever that it *had* to happen. Given that, the libertarian may well be unimpressed by the determinist's claim that the man's decision was relatively necessary. But here, of course, an advantage of this calculus appears: it exhibits in the very statement of the necessity what it is that necessitates the decision. Thus, that a decision is necessitated by itself is uninteresting and un-alarming; that it is necessitated by a preceding physical state *is* alarming. The difference is that in the first case what necessitates is not something which is (thus far shown to be) beyond the agent's control; in the second case that which necessitates is beyond the agent's control, and beyond even his knowledge.

(v) The fourth solution: elements of a new system

The prohibition on detachment in von Wright's system was, we suggested, a device for keeping explicit the admission that the necessity was relative. I wish to offer the elements of a system in which detach-ment is possible, and in which the distinction between absolute and relative necessity is marked by the use of different operators for each. These operators will be □ and L respectively. But rather than regard □ as necessity proper, and L as a kind of cheat necessity, a poor cousin, I shall regard L as full-fledged necessity proper, and □ as full-fledged necessity with an additional feature: that the necessity springs entirely from internal considerations – from the fact that the idea of the predicate is already contained in the subject, etc.

Thus Lp is to be read 'p is necessary', or 'p must be true'; □ p is to be read 'p is necessary from internal considerations alone' or, 'p must, from internal considerations alone, be true'.

Among the laws of this system will be:

(1) □p → LP
(2) LP → p
(3) □(p → q) → (□p → □q)
(4) L(p → q) → (Lp → Lq)
(5) L(p → q) → (p → Lq)[26] and hence also
(6) □(p → q) → (p → Lq)

(1) says that if p must, from internal considerations alone, be true, then p must be true. (2) says that if p must be true then p is true. (3) says that if, from internal considerations alone, p cannot be true unless q also is true, then if p must, from internal considerations alone, be true,

q also must, from internal considerations alone, be true. (4) says that if p cannot be true unless q also is true, then if p must be true, q also must be true. (5) is the useful and interesting new axiom: if p cannot be true unless q also is true, then, if p is true, q must be true. (6) is entailed by (5) together with (1); it says that if, from internal considerations alone, p cannot be true unless q also is true, then, if p is true, q must be true.

It is to be noted that

(7) $\Box(p \to q) \to (p \to \Box q)$

is not true and is not a law in this system.

Now the form of any law of nature will be $L(p \to q)$,[27] and so (5) permits an inference schema which will solve our problem:

(i) $L(p \to q)$

(ii) p

∴ (iii) Lq.

We have already made the point several times over that we do in fact reason this way about natural necessity. The resistance to admitting this inference schema into the standard modal logics has come from the fact that the parallel inference schema for *logical* necessity is not valid ((7) is not a law), and it is with logical necessity, necessity springing from internal considerations alone, that the standard modal logics have been concerned.

Of course, it is not only natural necessity which will be symbolized by L, but other sorts of relative necessity as well. The necessity which attaches to the conclusion of a valid argument when its premises are true is one of them (thus law (6)); another is the Aristotelian self-necessity to which we have already alluded.[28]

Now it might be thought that, on the grounds of what has just been admitted, this system is too strong: it admits too much as necessary. The libertarian will be unimpressed if the kind of necessity we succeed in ascribing to an event is a necessity which it has just in virtue of its actuality. But once again we reply that the important thing about the determinist's argument will be not just the necessity of its conclusion, but also from what givens that necessity can be derived. The givens of the argument are items which are beyond the control and even the knowledge of the agent; and yet the argument shows that those items, unknown to the agent, necessitate his decision. Thus, where P is a total physical state in and surrounding a man at t, and where D is his decision at t', to divorce his wife, the argument

(1) D

∴ (2) LD

though valid, is uninteresting to the libertarian; on the other hand the argument

   (3) L (P → D)

   (4) P

∴ (5) LD

which is also valid, is interesting, and provokes his concern.[29]

II

ON THE DISPUTE BETWEEN COMPATIBILISTS AND INCOMPATIBILISTS

In chapter I we saw the neurophysiological determinist battling against a formal difficulty which threatened to disarm his thesis, and we saw that there was every likelihood that he would win. Winning would consist in being able to show that any given decision of an agent was made necessary by a law of nature together with some initial conditions which were beyond that agent's control or even his knowledge. To put it another way, the neurophysiological determinist would be able to show that had a given decision not occurred as and when it did, a law of nature would have been violated. From there it seems but a short step to showing that, though the agent supposed that it was as much in his power to decide not to divorce his wife (say) as it was in his power to decide in favour of divorce, the agent was mistaken in this supposition. For in fact it was not in his power to decide against divorce - to decide against divorce would have been to violate a law of nature, and it was, of course, not in his power to violate a law of nature. And thus the determinist shows that we do not have the liberty of indifference in our decisions that we suppose we have: it is not really in our power to decide either way.

But many have denied that this short argumentative step is valid; they are compatibilists, arguing that one's having the ability to do (decide) otherwise is compatible with one's being causally determined not to do (decide) otherwise. The test for liberty of indifference in a decision is 'could he have decided otherwise?'; the compatibilist contends that the meaning of that question is such that it can properly be answered in the affirmative even when it was causally determined that he take the decision he took.

27

The literature surrounding this question is enormous, and in places the debate is, by modern standards, almost hot.[1] To review the whole discussion here would be tedious. What I propose is rather to enter the debate where it stands, with one small sharp counter-attack and one softer more grandiose sally. The first is an answer to a subtle and excellent compatibilist move, Kenny's argument from Scotus; in the course of answering this move I shall develop what now seems to me a knock-down argument for incompatibilism, though it works rather on the surface of our conceptual system and provides no real explanation of what the dispute is about. The more grandiose offensive tries to make good this deficiency; it is a reflection on language and meaning.

(i) On Kenny's Scotist argument

Kenny depicts the compatibilist as offering the following argument. In a case where it is physiologically determined that I not move my little finger:

I cannot violate a law of nature.

Moving my little finger is, in this case, violating a law of nature.

∴ I cannot move my little finger.

Indeed, we have already offered an argument very like this. But, Kenny objects:[2]

there is something wrong with the pattern of argument

I can (cannot) do X

Doing X is doing Y

∴ I can (cannot) do Y.

The principle

If to ϕ is to ψ, and I can ϕ, then I can ψ

which seems harmless enough is in fact false if it is considered as having unrestrictedly general application.

(The principle is true only when the identity between ϕing and ψing is logically necessary; in the incompatibilist's argument that identity is clearly contingent.) Kenny exhibits the fallacy of the principle with a counter-example: 'I may be able to hit the dartboard; on a particular occasion I may hit the dartboard by hitting the centre of the bull, but it by no means follows that I am capable of hitting the centre of the bull.'

Now before I reply to this impressive argument it must be made quite clear that the conclusions of the above argument forms have to do with ability on a particular occasion, not ability in general. The 'can' is what Nowell-Smith called the 'all-in' can,[3] such that 'I can ϕ now' means that I have both the ability and the opportunity to ϕ now. Even when this is made quite clear, of course, Kenny's counter-example holds good: from 'I can (in general) hit the dartboard', and 'on this occasion hitting the dartboard is hitting the bull', it does not follow that on this occasion I can hit the bull.

For in the logic of human ability *ab esse ad posse non valet consequentia*: from the fact that I *do* hit the bull it does not follow that I *can* hit it.

My reply to Kenny begins by suggesting that the logic of inability does not seem to mirror the logic of ability. What he has said about 'can' is perfectly correct, I think, but it does not seem to apply to 'cannot'; and since the incompatibilist's argument, as put forward by Kenny, uses 'cannot', Kenny's refutation of it is not secure.

The way in which the logic of 'cannot' seems different from that of 'can' is this. We have just enunciated the principle that 'I do ϕ' does not entail that I can ϕ. This principle seems to me quite unrestricted in its scope: I can think of no ϕ for which we should want to say that the entailment holds. But the contrapositive of this entailment does not seem to fail unrestrictedly: there are many ϕs such that 'I cannot ϕ' does seem to entail that I do not ϕ. For example: 'I cannot square the circle' seems to entail that I do not square it; 'I cannot lift six tons unaided' seems to entail that I do not lift six tons unaided. But of course there are also many ϕs for which this contrapositive entailment fails, for example: from the fact that I cannot execute a swan dive it does not follow that I do not execute a swan dive, for I may execute one accidentally, while frolicking; from the fact that I cannot hit the bull it does not follow that I do not hit it, for I may hit it by luck. The inference from 'I do ϕ' to 'I can ϕ' seems unrestrictedly invalid; but the contrapositive inference, from 'I cannot ϕ' to 'I do not ϕ', seems valid sometimes and sometimes invalid. We must say under what conditions it is valid and under what conditions it is not. Once we have done this we shall have shown that the two inferences are not really contrapositives – for contrapositives must have the same truth-value; we shall have shown, that is, that 'cannot' is not the logical negation of 'can'.

It is not difficult to arrive at a rough statement of the conditions under which 'I cannot ϕ' entails that I do not ϕ; the examples I gave suggest some. Thus, where ϕing is either logically or naturally impossible 'I cannot ϕ' seems to entail 'I do not ϕ'. 'I cannot make the hypoteneuse commensurate[5] entails that I do not; and 'I cannot stop the force of gravity' entails that I do not.[4]

We must analyse the startling result that 'cannot' is not the logical negation of 'can'. A normal claim that 'I can ϕ' is the claim, roughly, that I have the skill of ϕing; such a claim presupposes, though it does not state, that ϕing is logically and physically possible. By contrast the claim 'I cannot ϕ' can be used either to claim that I lack skill, as in 'I cannot ride a motorcycle', or to claim that ϕing is logically or physically impossible, in which case the question of having or lacking skill does not even arise. 'I cannot square the circle' does not make the claim that I lack the skill to do so.

'I can ϕ' makes a claim about a skill, and that claim has no logical relation with 'I do ϕ'. 'I cannot ϕ' makes either a claim about a skill or a claim about logical or physical impossibility; if it makes a claim about a skill it has no logical relation with 'I do ϕ', but if it makes a claim about logical or physical impossibility then it does have a logical relation with 'I do ϕ'. When 'I cannot ϕ' merely disclaims a skill, it does not follow that I do not ϕ; when 'I cannot ϕ' claims the logical or physical impossibility of ϕing, it does follow that I do not ϕ.

Put more simply, the asymmetry between 'I can ϕ' and 'I cannot ϕ' is that the latter is ambiguous, and only one of its two meanings is the negation of the former.[5]

Now what is the importance of this? The reason that the inference about being able to hit the bull failed was that the principle *ab esse ad posse valet consequentia* does not hold for human ability. We have found, by examining the contrapositive of this principle, that it does hold for human ability under certain conditions. Under those conditions, then, the inference schema impugned by Kenny will be valid. Thus, compare:

I cannot pole-vault.
Here, pole-vaulting is winning this bet.[6]
∴ I cannot win this bet.

with

I cannot lift two steamer-trunks.[7]
Here, lifting yours is lifting two (because I am already carrying mine).
∴ Here, I cannot lift yours.

The first inference seems invalid, because all I have to do to win the bet is to pole-vault – I don't have to be *able* to pole-vault; and from the fact that I cannot pole-vault it does not follow that I do (shall) not, for I may by accident or luck succeed in doing it. The second inference on the other hand seems valid. The first premise seems to claim that it is naturally impossible for me to lift two steamer-trunks (cf. 'that wooden platform cannot support the marble statue'); it follows that I *shall* not lift two steamer-trunks – neither luck nor accident can supply my inability in this case – and hence that I cannot lift your steamer-trunk in this case.

If I am right that the inference schema which Kenny impugns is valid when the 'I cannot' in the first premise is a claim of natural or logical impossibility, then the incompatibilist's argument is made good. His argument was:

I cannot violate a law of nature.

Here, moving my little finger is violating a law of nature.

∴ Here, I cannot move my little finger.

For the ϕ in question in the first premise is very conspicuously a case of natural impossibility. If there is a certain *invraisemblance* in the argument it can be removed by correcting the tenses to those in which the argument would normally be advanced:

I could not have violated a law of nature.

There, moving my little finger would have been violating a law of nature.

∴ There, I could not have moved my little finger.[8]

I conclude that Kenny has not shown that the incompatibilist's argument here fails. On the contrary, the argument seems incontestably valid.

(ii) Some considerations about meaning

The compatibilist holds that the meaning of 'I could have done otherwise' is such that that proposition can be true even if it was not naturally possible that I should have done otherwise. To claim 'I could have done otherwise' is to claim a number of things, but it is not to claim that natural determinism is false. Thus, for example, in Kenny's account,[9] to claim that I could have done otherwise is to claim that I had the ability and the opportunity to do otherwise; and one can of course truly be said to have the ability and the opportunity to ϕ at t, even if it is naturally determined that that ability shall not be exercised nor that opportunity taken, at t. To this, the incompatibilist will want to reply that while the latter part of it is true, it doesn't seem at all obvious that the meaning of 'I could have done otherwise' is exhausted by the conjunction 'I had the ability and the opportunity to do otherwise'.[10] The issue, then, is about the meaning of 'I could have done otherwise'. What does this mean? How are we to decide what it means? Pursued in this direction the compatibilism issue becomes an issue in the philosophy of language, and it is there that I should like to pursue it for a page or two.

I hope I shall not be guilty of erecting a straw man if I depict the compatibilist as pleading something like this: the *meaning* of a word or phrase is closely tied to its normal use; *it* means pretty much what *people* ordinarily use it to mean. Now people ordinarily use the phrase 'he could have done otherwise' to deny that certain fairly overt constraints were present when he acted: compulsion, duress, extreme moral pressure, and so forth. Ordinary people *never* use it to deny the thesis of universal determinism; that determinism is false is not part of the meaning of – or is not entailed by the truth of – 'he could have done otherwise'. Of course *philosophers* can mean what they like with the phrase, but if they are interested in *what the phrase means*, amid the ordinary human institutions, then they must look at what ordinary people use it to mean. And here philosophers *are* interested in what the phrase means amid the ordinary human institutions, for they are concerned to say whether and how those institutions must change if it is discovered that the sentence in question can never be uttered with truth.

Of course, anyone who is not a linguistic anarchist would want to agree that there is a difference between the meaning of a sentence and what a linguistically aberrant utterer of the sentence uses it to mean. But the compatibilist's plea looks as though it comes close to claiming that there is *no* difference between what a sentence means and what the ordinary utterer (in ordinary circumstances, etc.) uses it to mean. I think that this thesis is demonstrably false, and furthermore that there is a way of telling, among those things which are not meant in ordinary circumstances by ordinary utterers of a sentence, which are and which are not part of the meaning of that sentence.

The things one says often prove true in unexpected ways. Here is a very humble example. I open the door to greet a guest and see that the street is wet. I remark, 'Ah, either it has rained or the street washer has been by.' My guest replies, 'Well, I can tell you that both are true.' My remark has proved true in an unexpected way: I didn't intend my 'or' to be exclusive, and yet I hadn't it in mind that both disjuncts might be true, and I am surprised that it is so.[11] A more grandiose example is the forecast of the Pythia to Croesus that if he crossed the Halys he would destroy a mighty empire; the way in which that proved to be true was a surprise to Croesus, and perhaps to the Oracle as well.[12] A middle example might be this. I notice that a lake by which I am passing seems desolate and seems not to support any of the usual sort of plant life at the shore. I ask a local person if the water is polluted. He says yes, polluted by salt; large quantities of salt were put into it for research some years back. Salt is not normally considered a pollutant: we do not say the sea is polluted because it is salt; yet here being salt seems to be a way in which a lake can be polluted. It is a truth-condition for being polluted, but it is not one that I have in mind as I ask the question; I have in mind the ordinary sorts of pollutants: detergent, oil, sewage, factory effluent.

This common occurrence in language might be called the phenomenon of the *submerged truth-condition*. Our reflection on the meaning of 'I can ϕ' in the previous section drove us to distinguish, among the truth-conditions of a remark, the one that was, so to speak, the claim that was being made, from the others which were not being claimed but *presupposed*. Thus in saying 'I can speak Russian' I am claiming a skill, and presupposing the natural and logical possibility of the feat. But presuppositions are only one kind of submerged truth-condition; they are the submerged truth-conditions which are necessary conditions. There is also a whole class of non-necessary submerged truth-conditions; in the normal case any one of these, taken together with the presuppositions, yields sufficient conditions for truth.

Typically, this phenomenon arises with relative, or incomplete, concepts. 'Polluted' is incomplete and is to be completed as 'polluted by x'. 'Married' is incomplete, and is to be completed as 'married to x'. And so forth. The *meaning* of such a predicate as 'polluted' is to be regarded as the list of all the necessary conditions for its ascription, together with the list of all the non-necessary conditions for its ascription which, taken together with the necessary ones, will yield sufficient conditions for truth. The *meaning*, then, of predicates is a fairly long list, partly disjunctive.

On any one occasion of the ascription of the predicate, one of these conditions – necessary or non-necessary – will be the claim that is being made. On any occasion of asking about the ascription of the predicate, one of these conditions, or a typical range of them, will be the conditions about whose fulfillment one is inquiring . But in any case the *meaning* – the whole list of necessary and non-necessary conditions for truth – far outstrips what is being meant, the claim that is being made or asked about. All the conditions which are not the claims being made or asked about are on that occasion submerged. The things one says can prove true in surprising ways.

But, in the case of the non-necessary conditions, how are we to tell whether an alleged one is or isn't really in the list, is or isn't part of the meaning of the term? We can begin by looking at a case where it isn't. Suppose I say of a woman whom I have met very briefly, 'I would say she is married.' My interlocutor, who is better informed, says, 'Yes, she is married – to Christ; she is a nun.' Here (if the matter is anything more than idle gossip) I am likely to object and say that that didn't count as being 'married'. That is, nuns are married only in a rather special sense of 'married': being married to Christ isn't a way of being married at all, in the ordinary sense of that word (compare here: being free from determinism isn't a way of being free in the ordinary sense of that word). In this case, then, we might say that my interlocutor has alleged a submerged truth-condition for my remark; but I have rejected the allegation.

How do we tell the difference between these two cases? How do we tell the difference between a submerged truth-condition of a remark and something which isn't a truth-condition of that remark at all? I suggest that we can tell the difference fairly securely by watching the linguistic behaviour of an ordinary person to whom the submerged truth-condition is suggested. If he says at once, 'that's not what I meant by –', then we can take it that the alleged submerged truth-condition isn't a truth-condition of the ordinary concept – ; if, on the other hand he reacts with interest and surprise, we can take it that what was offered was indeed a submerged truth-condition of the ordinary concept – . It is the unreflective behaviour of ordinary speakers which will supply the best evidence for the submerged content of ordinary concepts.

Well, the incompatibilist holds that the truth of universal determinism is a submerged truth-condition of 'he could not have done otherwise', and the compatibilist holds that it is *not* a truth-condition of that remark, in its ordinary sense, at all. Can the matter be put to the test? Clearly, it can, and very often is. Every time some or other thesis of determinism is expounded to students in their first year of philosophy, this matter is put to the test. And clearly it is the incompatibilist who wins by this test: students do not typically (in my experience, never) react with scorn and derision, but with something much more like shock. There is nothing in the behaviour of ordinary speakers to suggest that universal determinism is not really a truth-condition for 'he could not have done otherwise'. And what better test could there be of the limits of ordinary concepts than this of trying what ordinary people unreflectively admit and what they exclude from those concepts?

III

THE CORRELATION THESIS

In the introductory chapter the Correlation thesis was stated rather crudely; the present chapter will be devoted to some speculative refinements upon that statement.

Grosso modo we know that there is a correlation between brain event kinds and mental event kinds. We know, for example, from Penfield's famous experiments, that when a certain bit of the occipital lobe of the cortex is stimulated electrically, a conscious patient 'sees' a flashing red light. We know from the experiments of Hubel and Wiesel at Harvard and from work done in their wake that individual cortical cells are responsible for the seeing of individual points in the visual field.[1] We know from the clinical work of Russell in Oxford, for example, that lesions in different parts of the brain result in various deficiencies of memory.[2] We know from the work of Brenda Milner in Montreal that injuries in certain other parts of the brain result in certain deficiencies of hearing.[3] And from the work of A. R. Luria in Moscow we begin to have a general picture of what part of the brain is responsible for what mental skill or ability.[4]

But the sort of Correlation thesis that is verified by these findings is a very crude one, much too crude to serve the neurophysiological determinist's turn. What the determinist wants is a very fine-grained Correlation thesis to be true; he wants it to be true that to every mental event kind there is correlated a neural event kind, where the mental event kinds in question are *infimae species* – not the much broader kinds in which neuropsychology, as we know it now, deals. We shall have to become clear what we mean by *infimae species* of mentals; that will be our concern in section (iv) of this chapter; but first there

36

are some preliminary matters to be sorted out.

(i) Two objections to a Correlation thesis

To the proposal that every mental event kind is correlated to a neural event kind down to the finest grain two very general kinds of objection have been advanced. These objections are deep and powerful, but though they would preclude various versions of the Correlation thesis, there is one important version which they would not preclude, and which is the version the determinist requires to make his case. I call these two objections the thesis of the Extramentality of the Mental, and the Taxonomic Incongruence thesis.

The Extramentality of the Mental
Those who believe in the possibility of establishing fine-grained correlations between mental events and brain events are often in the grip of an old-fashioned view about the mind, according to which mental events are private happenings in a person's consciousness, whose nature it is to have actually precisely the qualities they appear to have, to occupy space in an odd way, and time in the usual way. From the time of Descartes, indeed, until that, perhaps, of Wittgenstein, this view of the mental has prevailed, giving rise both to the familiar battery of sceptical problems and to the mind-body problem. Such a view of the mental clearly invites the formulation of a comprehensive Correlation thesis: according to it the realm of the mental is exhausted by this train of private phantasms or occurrent conscious states, and so a Correlation thesis would neatly tie each mental item to a cerebral one.

Since Wittgenstein a number of objections to this traditional view of the mental have been advanced which might summarily be described as pointing out the extramentality of the mental. By that I mean that they have shown that the truth-conditions for the ascription of many mental predicates (that is, predicates which we should intuitively agree are mental) consist in part, largely, or even wholly, of items other than occurrent conscious states. The realm of the mental is thus much wider than we used to think.

For example, we should agree that 'abashed' is a mental predicate, but a person cannot properly be called abashed unless he or she exhibits a certain form of behaviour; 'acquiescent' is a mental predicate, but a person cannot properly be called acquiescent unless he or she has a certain history of a pattern of behaviour; no one can be abusive, acerbic, acrimonious, or adamant without a certain kind of linguistic behaviour and without the public institution of settled meanings of words; I cannot absolve you from your debts unless there was a public institution of debt-acquisition behaviour and, historically, you behaved in such a way as, under that institution, to acquire debts to me. And so forth. It turns out to be the rule rather than the exception that the truth-conditions for the ascription of mental predicates involve extramental items; the ones of which we have so far seen examples are: (a) present behaviour of the subject, (b) past behaviour of the subject, (c) public institutions of meanings etc., (d) another person's past behaviour. Extension of this list would lead us to include (e) future behaviour, (f) subconscious states, (g) past or future conscious states. (The latter are not quite properly extramental.)

Extreme versions of this claim have held that *all* the truth-conditions for the ascription of mental predicates are extramental in this way; more moderate versions hold only that some of them are. What at any rate emerges indisputably is that the meaning of most mental predicates is very far from exhausted by mental events of the traditional sort, conscious states.

Now this observation about the extramental strands of meaning in most mental terms tends against the Correlation thesis in the following way. While we might manage to think that there are lawlike correlations between brain states and conscious states, it is simply inconceivable that there should be any correlation between neural states of mine and the existence of public institutions, or other persons' behaviour in past, present, or future time. One can perhaps imagine that the nervous system is the machinery which subtends conscious states, but it can hardly be the machinery which subtends all sorts of complicated 'non-mental' states of the world. One might have hoped for a lawlike correlation between conscious states and neural states; mental states prove to be much broader-based and more complex entities than mere conscious states; it is impossible to think that their non-conscious ingredients are correlated with anyone's neural states, and so it is impossible to think that *they* are correlated with anyone's neural states. So that while correlation between conscious states and neural states is a possibility, correlation between mental states and neural states is not.

It is indeed instructive to leaf through a small dictionary and see how very very few words which we should intuitively allow to be mental are such that mere conscious states could amount to sufficient truth-conditions for their ascription.[5] Even feeling words like 'sadness', 'remorse', and 'shame' seem, on close inspection, to make essential reference to a state of affairs in the world; the raw feelings involved in each do not obviously differ. In the end we are left with a short list of 'raw feel' words – a list no doubt inadequate to catch with any subtlety the range of raw feels that can in fact occur – and very little else. One curious upshot of this is the following. It is often said that the criterion of the mental is incorrigibility, that a person cannot be mistaken about the mental state he or she is in. At best, though, incorrigibility is the criterion of conscious states, and it turns out that the set of predicates which describe pure conscious states is so impoverished that incorrigibility about them is purchased at the price of having virtually nothing to say about them.

The thesis of the Extramentality of the Mental seems, then, to threaten the Correlation thesis in two ways, in a principal way and in a corollary way. Since most mental terms are such that some of their necessary truth-conditions are not conscious states, there seems to be no hope of correlating all or even many mental states with neural states. Moreover, if we accept the corollary of the Extramentality thesis that the few conscious state terms that we do have are very inadequate to the subtle differences among our conscious states, then such a mental-neural correlation chart as we might produce would be a very short and summary affair, not nearly rich enough to serve the determinist.

The neurophysiological determinist's strategy of reply to all of this has two parts. The first is a small revision to the Extramentality thesis, and the second is the realization that it and he are really discussing different things.

In the first place, then, we must not, in our enthusiasm for extramentality, miss the point that many mental terms which seem to involve essentially some reference to external, nonconscious, states, do not in fact do so. We said that 'sadness', 'remorse', and 'shame' did not differ obviously in the raw feelings they involve but rather in the state of affairs in the world to which each is a response. But that is overhasty. What makes sadness sadness and remorse remorse is not that in the one case something regrettable for which I was not to blame happened whereas in the other I did a blameworthy thing, but rather that in the second case *I believe* I was to blame and in the first not. That is, it is often not external states of affairs themselves but rather my beliefs about external states of affairs which make my mental states what they are. I can have degraded myself utterly, but unless I believe that I did so I cannot feel shame. And belief in these cases is a conscious state. So that a good many mental terms are thus rescued from extramentality. An important one among these, for the neurophysiological determinist, is decisions. I can genuinely decide to divorce my wife even if my belief that I have a wife is mistaken. To be sure, I shall have trouble genuinely carrying out the decision, but the decision itself is genuine.

But secondly and more importantly the determinist must reply to the Extramentality thesis that it really is beside the point for him. For in fact he has no particular interest in matching up all the mental predicates in the dictionary with neural descriptions. What he wants to match up with neural states is conscious states – or neural descriptions with descriptions of conscious states. To be sure, it is an annoyance for him that the mental language does not describe conscious states: that so much mental language describes other things, and that the mental language which does describe conscious states is desperately crude. The burden that the Extramentality thesis lays upon him is that of devising a language which will designate conscious states in their purity, and which will do so in a way that is not crude. In a later section of this chapter we shall see how he can discharge this burden.

The determinist's escape from the Extramentality thesis can be summarized in this way. The Extramentality thesis is a thesis about the mental language and its imperfect relation to conscious states. The determinist wants to discuss only conscious states, and is indifferent to the nature of the mental language. Theses, therefore, about the oddness of that language pass him by. In what follows I shall speak a good deal of mental events and mental states, and by them I in every case understand conscious events and states. That these cannot straightforwardly be captured by the mental predicates of our language has no bearing on the argument.

The Taxonomic Incongruence thesis

One implication of the truth of the Correlation thesis is that it would in principle be possible to draw up what might be called a fine-grained correlation table, matching mental event kinds with neural event kinds. What would such a table be like? We can imagine that on the left side it would list the neural event kinds and on the right side the mental event kinds which correspond. (Of course, a good many of the spaces on the right hand side will be blank, for many neural event kinds are correlated with unconsciousness or dreamless sleep – a lack of mental events.)

Of course, constructing such a table would presuppose that we had some way of designating mental events. We have already seen that the ordinary mental language speaks of much that is extramental. We might have thought that a simple way to designate mental events (i.e. conscious states) would be to devise a *conscious-state-extractor* operator to the normal mental language. Its symbol might be the letter '*M*' prefixed to an ordinary mental word. Thus whereas the truth-conditions for 'abashed' involve both some conscious states and some behaviour, the truth-conditions for '*M*abashed' would be only the occurrence of the conscious states which are part of the truth-conditions for 'abashed'. The *M*-prefix extracts only the conscious states from the assorted ingredients in the meaning of the mental term to which it is attached. (The right side, then, of the correlation table would consist first of some bona-fide raw feel words, and then a long list of *M*-prefixed mental words.) In this way we could employ something closely resembling the ordinary mental language to designate mental events on our correlation table. We should be classifying conscious states on the grid of the ordinary mental language.

Here at once we shall encounter an objection from adherents to what might be called the Taxonomic Incongruence principle: that the ways physiologists have of classifying neural states and events are not (likely to be) mappable onto the ways that psychologists and ordinary people have of classifying mental events. This principle is almost certainly true. Luria, for example, has confirmed it strikingly with his demonstrations of the way in which a single mental predicate matches up with what he calls a 'constellation' of neurophysiological happenings.[6] No sober person can be surprised by this principle.

Davidson in his famous paper 'Mental events'[7] is, I believe, reflecting precisely this principle of Taxonomic Incongruity when he proposes 'anomalous monism', the view that though every mental event is identical with some physical event, and though mental events interact causally with the physical world, there are no psychophysical laws. The reason for this startling idea is that whereas causality and identity are relations between, so to speak, brute events, laws are relations between descriptions of events: laws are linguistic. So that although mental events, in their capacity as the physical events with which they are identical, can have causal interactions with the physical world, these causal interactions cannot be framed as laws. Psychophysical laws cannot be stated because the mental language cannot be mapped onto the physical (i.e. neurophysiological) language: the two languages do not cut up the same reality into slices which correspond. To try for a law relating physical effects to mental causes is like putting on mittens to play the piano.

A more elaborate but more accurate analogy is this. The academic dress of the University of Oxford is a science in itself, but one simple feature of it is very nearly constant: the robes worn signify degrees held and not office in the University. There is no special robe for the vice-chancellor; there are no special robes for professors; there is none for the registrar; and so on. Now academics can be classified by the degree they have, or by the position they hold in the University. These two systems can be thought of as two languages, one physical and the other mental, in the analogy. Trying to make psychophysical laws is like going to an academic ceremony and whiling the time away by trying to match up robes and offices in a lawlike way: it cannot be done. Robes match degrees held (physical interacts with physical); and degrees held do not correlate with office. The vice-chancellor may well be an MA while the sub-registrar has an exalted doctorate.

The psychophysical laws which are precluded in Davidson's scheme are of two sorts: laws relating the mental description of an event to the physical description of the same event; and, *a fortiori*, laws relating the mental description of an event to the physical description of some *other* event, linked causally to the first. The first sort of psychophysical law sounds identical to what we have called the Correlation thesis. The two are however not identical, and indeed I want to reject the one and accept the other. I want to agree with Davidson that there are no psychophysical laws and also to defend the Correlation thesis. Davidson's position that there are no psychophysical laws seems to me to amount to this: the mental language as we now have it does not map onto the physical language as we now have it, and so. . . . With this I entirely agree. The Correlation thesis I envision claims rather that there is a way in which the mental can be classified, and there is a way in which the neurophysiological can be classified, such that to every mental event kind there is correlated a neurophysiological event kind. That is, while admitting that taxonomic incongruence makes it impossible to map *our* mental language (as revised, of course, by the universal use of the *M*-operator) onto *our* neurophysiological language, I do not think that there cannot be produced a mental language and a neurophysiological language which would map onto each other.

If we admit that laws are linguistic in the sense that they link not events but descriptions of events, then we immediately start up a rather tedious debate about what it is for laws to exist: is it for there to have been elaborated some descriptions by which the link can be effected, or is it just for some such descriptions to be elaborable? I don't care which we say; but if the first then there are no psycho-physical laws, and if the second then there are. The incompatibilist is quite as worried by the thought that laws can be made as he would be by the thought that they have been made, of course, and so nothing in the Taxonomic Incongruence principle or in Davidson's anomalous monism is of any comfort to him: no important version of the Correlation thesis is undermined.

So, to make good my defence of the Correlation thesis against the Taxonomic Incongruence principle, I have to get enmeshed in language making. I have to make up description systems for the mental and the neurophysiological which will map onto each other and which (of course) will permit causal laws to be framed linking the neurophysio-logical to the rest of the physical. Harré, facing this difficulty which stands in the way of a science of neuropsychology, recommended what he called the taxonomic priority of the mental: that the ordinary mental language should stand and the neurophysiological language be revised to conform to it.[8]

However, we cannot get off quite so easily, for a delicate circum-stance requires that we use neither the ordinary classifications of physiology nor the ordinary classifications of psychology in setting them up – at least not to start with. The delicate circumstance will be treated in section (v) below, but it can be summarized *ambulando* as follows: we are assuming that the kind of mental state I am in at t is correlated with the kind of neural state I am in at t, but we cannot securely infer from this that a certain *part* of my mental state kind, e.g. feeling ill-at-ease, is correlated with a certain *part* of my neural state kind.[9] Now the classificatory systems both in psychology and in physiology deal with parts or aspects of the whole state of mind or whole state of body at a given time. We must not use these natural classificatory systems, therefore; we must devise classificatory systems which are capable of describing the whole state of the mind, and the relevant features of the whole state of the brain, at a time. We know (assuming the truth of the Correlation thesis) that whole mind state kinds are matched with whole brain state kinds; but it is not clear that we can infer from this that parts are matched with parts. For example, for the sake of argument let this be a complete description of my whole state of mind at t: seeing the vice-rector carrying my dossier, against the background of a neo-classical building, with trees about, hearing traffic sounds, feeling hot, feeling ill-at-ease. Call this mental state kind M1. Assuming the Correlation thesis, we know that M1 is correlated with a certain neural state kind N1. But we don't know that some little bit of M1 – say, feeling ill-at-ease – has its own little representation in the total description N1. We don't know that if on another occasion my mental state was very different from M1 but had in common with it that I felt ill-at-ease, my neural state then would have a little bit in common with N1, the bit which represents feeling ill-at-ease.

For this reason we must lay aside the ordinary mental language (revised), which designates only parts of whole mental states, and we must lay aside the ordinary neural language, which designates only parts of whole states of the brain. We must devise in their stead systems of identifying whole brain states and whole mental states, *infimae species* of neurals and *infimae species* of mentals. Such *infimae species* of mind and brain states are what we must seek to correlate on our correlation tables.

The two major objections to a Correlation thesis are thus circumvented. We avoid the embarrassments of the Extramentality thesis by realizing that we really are not interested in correlating with brain states anything but conscious states; that the mental language is a net which picks up a lot of other fish as well is only a small annoyance to this interest. We avoid the more serious embarrassments of the Taxonomic Incongruence thesis when we see that we must in any case discard both the classificatory system for conscious states that stems from the ordinary mental language, and also the classificatory system for brain states that neurophysiologists use. That these two systems are not congruent with each other is therefore a matter of indifference to us in making out the neurophysiological determinist's case. We are however committed to making up a classification system for *infimae species* of brain states and of mind states, and to that task we shall turn in section (iii) below. First we must notice a structural feature that the determinist will require of the correlation table, and we shall see an argument for believing that his requirement can be met.

(ii) One-one or one-many, but not many-one

The statement of the Correlation thesis must be of the form 'to every mental event kind . . .', not 'to every neural event kind . . .'; for there are many neural event kinds which have no mental event kinds correlated with them: the ones that go on in dreamless sleep for example (dreamless sleep is not a mental state but the absence of mental states), or in the state of coma. If we depict the stream of mentals and neurals in a man's life using the conventions of the diagrams in chapter I, we shall have the situation shown in Figure 3.1 where only some of the neurals will have mentals correlated with them.

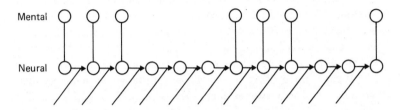

Figure 3.1 One-one or one-many but not many-one

And, that said, the thesis may be of two forms, one-one or one-many. That is, the thesis will be strong enough for the neurophysiological determinist whether a given mental event kind can be correlated with only one neural event kind, or with several neural event kinds. To put it another way, it doesn't matter to the determinist thesis whether a given mental event kind can be mediated by only one or by several different states of the neural apparatus. As long as the thesis permits the inference from the occurrence of a given neural event kind to the simultaneous occurrence of one only mental event kind, the neurophysiological determinist's claim is accommodated.

What the thesis must not be is many-one; that is, it must not be the case that from the occurrence of a given neural event kind one can infer the occurrence of one out of a disjunctive list of mental event kinds. Suppose that neural event kind N1 is correlated with the disjunction of mental event kinds M1 or M2. Then from the occurrence in a man of N1 all one could infer would be that either M1 or M2 occurred. Now M1 might be (or include) the decision to divorce one's wife, and M2 might be (or include) the decision not to divorce one's wife. If that is the arrangement, then the determinist may be able to argue that N1 had to occur, but he cannot go on to argue that therefore M1, say, had to occur – for M2 could have occurred instead. That is, all the determinist could claim would be that it was necessary at the time N1 occurred that the man either decide to divorce his wife or decide not to divorce his wife (not a disjunction of necessities, but a necessary disjunction). And that is a degree of determinism at which few would balk.

The neurophysiological determinist, then, requires that the mental-neural correlation thesis be one-one or one-many, and that it not be many-one. Is there any reason to believe that reality matches his requirements of it? Is there any reason to believe that the sort of correlation which exists between mental kinds and neural kinds is not many-one? There is a sort of speculative reason to believe this, or at least to believe that if there is any discipline at all in the correlation of mental and neural kinds, then the correlation is not many-one. It is very far short of being a knock-down argument, but it does at least make it seem more likely that there is not many-one matching than that there is. The reason is that there is probably an apparatus for having colossally many times more kinds of mental events than a man can possibly have in his lifetime – about three thousand million times more.

Here is the calculation which leads to this conclusion. There are about 10^{10} neurons in the human nervous system. At any one instant each of them may be firing or not firing – ON or OFF. That means that the total possible number of whole neural state kinds in a man for any one instant is about two to the power 10^{10}, or ten to the power 3,000,000,000 – assuming that if we have listed each neuron and said of it whether it is ON or OFF we have given a total neural state description. Now, the highest frequency of firing in human neurons is about 5,000 times per second, so that we can say that ten to the power 3,000,000,000 is the number of different possible total neural states of 0.0002 second's duration. The number of such 0.0002 seconds lapses in a life of about a century's length is (100 × 365 × 24 × 60 × 60 × 5000) approximately 3.6×10^{13}. Thus if a man lived one hundred years, having a completely new total mental state kind every 0.0002 second, he would have used only a minuscule fraction of the total possible number of his neural state kinds.[10]

Of course there are a good many assumptions built into the above argument, and no doubt some of them will want revising, but it is unlikely that the revisions will have serious impact on the enormous disparity of these numbers, 3,000,000,000 orders apart. Given these figures it would be an unwarrantable frugality in nature to double up on the assigning of mental roles to each neural kind.

(iii) Designating neural kinds

A system for designating whole brain states[11] is in fact not very difficult to make, if we assume one thing. I wrote above that we wanted to describe the whole state of the mind, and the relevant features of the whole state of the brain at a time. I shall assume that the relevant features of the whole state of the brain are the electrical ones. Of course there are other features of the brain – chemical ones, for example – but these seem subservient to the electrical ones: adrenalin, for example, is a transmitter substance in the sympathetic nervous system;[12] its function is to transmit 'charges' from one nerve cell to the next. I shall assume, that is, that what counts for the correlation is the electrical state of affairs in the brain; how it got to be what it is, with the aid of what chemicals or what outside interference, doesn't matter. I shall say no more in defence of this assumption than that it seems to be usual.

This assumption allows us to envisage a scheme, adumbrated in the previous section, for distinguishing total brain state kinds. Since a neuron is either firing or not firing at a given instant, it would seem that we can exhaustively describe the electrical state of the brain at an instant by listing all the neurons in the brain (a longish list) and saying of each whether it is ON or OFF. There will then be $2^{(10^{10})}$ different possible instantaneous neural state kinds. Since the highest frequency of firing in the human nervous system is 5,000 cycles per second, we can say that there will be $2^{(10^{10})}$ possible neural state kinds of 0.0002 second's duration. Now we might want to say that a mental event as short as 0.0002 second is scarcely conceivable; we might want to place a lower limit on the possible duration of a mental event. If we say, for example, that the shortest conceivable mental events are one-twentieth of a second long, in each such one-twentieth-second period there will have been 250 successive 'instantaneous' brain states. If we then say that for the purposes of the correlation table a brain state kind has a duration of one-twentieth of a second[13] – call this an extended brain state kind – the number of different possible such extended brain state kinds will be the number of objects raised to the power of the number of slots:

$$(2^{(10^{10})})^{250}$$

That is the number of extended brain state kinds that are logically possible; no doubt a certain number of these will be excluded by the constraints of natural possibility. We can then give each one of these extended neural state kinds a number, and use that number to designate the extended neural state kind on our correlation table.

We have then a mathematically neat, if practically cumbrous, method of designating brain state kinds. After two short digressions to notice implications of the foregoing, we shall go on, in the next section, to the traditionally insoluble problem of identifying mental state kinds, and we shall see that there is an easy way of solving it as far as the neurophysiological determinist thesis requires it to be solved.

First digression: in the fine grain one man's nervous system cannot be mapped onto another man's nervous system. It is a direct consequence of this that the sort of correlation table we envisage will apply only to one man; there will have to be a different one for each man – though no doubt they will display some broad similarities, like thumbprints. If this is dismaying, it is at least not surprising. Even gross correlation tables are not universal: when part of the brain is damaged other parts may take over the function which naturally belonged to the damaged part.

Second digression: it is normally supposed that the electrical activity which is directly relevant to mental phenomena is that of the *central* nervous system (CNS) or brain. If we adopt this hypothesis we shall have to make some small changes in our diagrammatic representation of the neurophysiological determinist's view of what goes on in the mind and brain. Figure 1.2 will give way to Figure 3.2, reflecting this change.

But even Figure 3.2 is not the full and final picture. Events in the CNS also have direct causal effects on other neural events; this additional point is depicted in Figure 3.3.

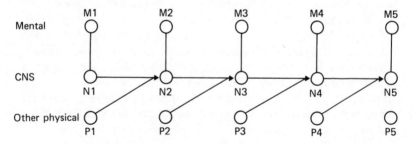

Figure 3.2 Neurophysiological determinism:
third approximation

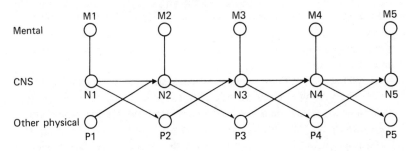

*Figure 3.3 Neurophysiological determinism:
fourth approximation*

Unless we allow the 'other physical events and states' to cover the whole physical universe or a causally isolated part of it there will be no licence for joining P1 to P2 with an arrow, alleging that P2 is uniquely determined by N1 together with P1.

(iv) Designating mental kinds

It is easily seen that exhaustively to describe a total mental state of one-twentieth of a second's duration is an undertaking well-nigh unthinkable, even for the skill and tenacity of a Proust or a Joyce. For one thing, most of our mental vocabulary is built up on mental events which must last more than one-twentieth of a second; for another, the reporting of the precise nuances would take many words and a long time; for another, the mentalistic vocabulary is often insufferably imprecise. Indeed it is so imprecise that we should almost certainly find that many of the subtler distinctions one would want to make in one's mental states rely on a somewhat idiosyncratic use of words which have little or no communicative effect in the public language; one would fall back on the wholly admissible remark: 'Well, *I* know what I mean by this distinction.'[14]

But if that sort of retreat to a private bit of language is pretty well inevitable anyway, why not avoid all the Joycean descriptive effort and simply set up a private designation system? The system I propose is remarkably simple. It is to designate a total mental state kind by mentioning the time at which it (first) occurred in the life of the person whose mental state it was. It must quickly be emphasized that this is not a technique for numerical individuation, but for qualitative individuation - designation of types, not of individuals. In humbler, looser, ways it is a technique we use often enough: 'It's going to be a 1968 kind of summer'; 'We're having a Quattrocento sort of renaissance here', and so forth.

Of course, this is in no way a practical system for designating mental kinds. The man who could use it on himself would have to have a quite phenomenally good memory, and one eye always on the clock. But then this whole scheme of correlation tables is of no practical interest whatsoever. After all, if a man recorded the state of one neuron every second, and never slept, it would take him three centuries just to write the description of his whole neural state at one instant; this hasn't the makings of a practical scheme. But then the scheme isn't intended to be practically, only theoretically, of interest: we are imagining correlation tables in order to see what would be entailed by the fine-grained Correlation thesis which the neurophysiological determinist requires. The fact that no human being could know these entailed details to be true does nothing, of course, to assuage the fear that they may be true.

With the technique of designating mental state kinds for a given person by mentioning the time at which they first occurred for that person we have reached the *infimae species* of which we spoke above.[15] Total mental state kinds, then, can be represented by numbers; and we earlier found a way of designating total neural state kinds by a system of numbers. The correlation table for a given man, then, can be thought of as two parallel lists of numbers whose members are appropriately matched.

(v) The 'style' of correlation: additive versus non-additive

The correlation of *infimae species* of mentals with total neural state kinds which we have just described can be thought of as providing the basic correlation table for a given person; the version of the Correlation thesis which entails that such basic tables can be made might be called the basic Correlation thesis. The basic Correlation thesis will certainly serve the purpose required of it by the neurophysiological determinist. But it is an unwieldy thesis, and it has the embarrassment that it is utterly incapable of verification by man - at least so long as his capacities are as limited as they are now. We might wonder therefore if a less unwieldy and perhaps even a useful correlation thesis can be extracted from it, one which instead of correlating these troublesome *infimae species* of mentals and neurals, correlates instead rather more abstracted kinds: parts or aspects of total mental state kinds with parts or aspects of total neural state kinds. Might one, for example, find that the partial mental state kind 'deciding to divorce one's wife' is correlated with such and such a partial neural kind, say, the firing of neurons 1 and 2? That is, might it be the case that whenever neurons 1 and 2 are firing one is deciding to divorce one's wife – no matter what is going on in the rest of the nervous system?

Well, this might be the case. It depends on whether the style of correlation is generally additive or generally non-additive. The following highly simplistic example will explain the difference between these two styles.

Let me, for the sake of argument, assume that I have sixteen nerve cells in my CNS. My CNS can then be represented diagrammatically as in Figure 3.4. Each of the nerve cells has two possible states, ON or OFF; in the diagram, ON is represented by shading in the cell, and OFF is without shading. Thus the whole state of my CNS at any moment can be represented by one of these diagrams, and 2^{16} different such diagrams can be made: there will be 2^{16} possible different states of my CNS.

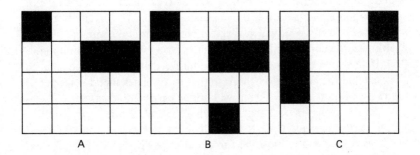

A   B   C

Figure 3.4 The 'style' of correlation

Suppose, now, that A is the diagram of the state of my CNS correlated with the mental event of seeing a robin – or rather seeming to see a robin. The question arises whether the neural correlate of that mental event consists (a) in the list of neurons that are ON, or (b) in the list of neurons that are ON and the list of neurons that are OFF. At first it may seem that there is no real question here, since a list of the ON neurons entails a list of the OFF neurons: any neuron which is not

ON is OFF. But the matter is more subtle than this.

What I have in mind is that the list of neurons that are OFF could be the correlate of the experience in two ways: (a) their being OFF could be the correlate of the experience's being no more than it is; or (b) their being OFF could be the correlate of the experience's being what it is at all. In case (a) if some OFF neurons are put ON, the experience will still be that of seeing a robin, but something else will be added, hearing a bell, say. In case (b) if any OFF neurons are put ON then the new experience will not be 'seeing a robin + − ', but something entirely different, having an itch in the toe, say. The style of correlation exhibited in case (a) we may call *additive correlation*; in case (b), *non-additive correlation*. If the style of correlation is additive, then the OFF neurons are the correlates of the experience only in an attenuated sense; they are the correlate of the experience's having no more content than it has; if the style is non-additive then the OFF neurons are the correlate of the experience in a much fuller sense.

Referring again to Figure 3.4, where A is the diagram of the neural correlate of seeing a robin; if the style of correlation is additive, then B will be the correlate of seeing a robin and (say) hearing a bell; if the style of correlation is non-additive, then B will be the correlate of some quite other experience, an itch in the toe (say). Further, if the style of correlation is non-additive, the correlate of seeing a robin and hearing a bell might well be C, which diagram has no parts in common with A, although the corresponding mental states have something in common.

If the style of the correlation is additive then it will be possible to streamline the Correlation thesis so that it correlates superior species; if the style is non-additive then it will not be possible to do this.

And it seems in fact that the style of correlation is partly additive and partly non-additive. It is additive to the extent that different modes of perception have their own quite specific areas of the cortex, so that neural activity there is correlated with that perceptual awareness; it is non-additive to the extent that without a certain general level of neural activity in the CNS there is no consciousness at all: Penfield's experiment doesn't work on anaesthetized patients. But it is not now possible to say anything about the enormous ground between these two opposing pieces of evidence.

For the purposes, then, of advancing the thesis of neurophysiological determinism it will be best to stay with the basic Correlation thesis.

(vi) How correlations are established

The foregoing considerations prompt an intriguing question, one which could perhaps be answered experimentally. And what seems to me the likeliest answer to the question suggests a way of solving the free will problem - or anyway of refuting determinism - on condition that there is no or little correlation of superior species.

In his book *A Materialist Theory of the Mind*, D. M. Armstrong offers what he takes to be a conclusive argument against behaviourism.[16] It is that if one produced a metal replica of the CNS of a given man, and stimulated the outermost of its afferent 'nerves' electrically in just the same pattern that the outermost cells of the afferent nerves of the CNS of that man had been stimulated throughout his life, and if one similarly arranged to use up the energy coming from the efferent 'nerves', then at any given time that metal replica would be in exactly the same total electrical state as was the CNS of the man at the corresponding time. Clearly then one would have to say that the replica was in exactly the same (kind of) *mental* state at that time as was the man at the corresponding time.[17] But the metal replica would not exhibit any behaviour: it is just a bunch of wires in a box. Therefore mental states can be attributed to an object in cases where there is no behaviour; therefore behaviourism is false.

I am not sure how daunting a behaviourist, if there is one, would find this argument to be; but for us, anyway, its interesting step is the second one, the assumption that wherever a certain pattern of electrical excitation occurs there also occurs a certain mental state kind. This assumption might be called the hypothesis of Universal Preordination of Correlates. There could be various forms of this hypothesis: one, for example, might add the constraint that the electrical state must be in neurons, not in inorganic matter; another might hold to an Individual Preordination hypothesis, saying that one's correlation table, like one's thumbprint, is in one's genes. We needn't however be worried about these disputes here. One of the implications of this hypothesis is that to the extent that men's nervous systems differ, men are incapable of having qualitatively the same mental state; but that, I suppose, is not an embarrassing implication. What is mysterious about the Universal Preordination hypothesis is *why* this or that electrical pattern should be preordained to be correlated with, for example, seeming to see a red patch - what have electrical charges to do with redness? But this is the sort of philosophical mystery that goes in and out of focus; sometimes it seems intoxicating and vertiginous, and sometimes it seems merely one of the world's facts, like the fact that yellow light has a frequency of 5.2 × 10¹⁴ cycles per second. On this philosophical terrain we must expect to meet that kind of mystery.

One can however imagine a competing hypothesis with this one, an hypothesis which holds that there is no preordained correlation but that what goes on is something like this: before a man is conscious no correlations are ordained. His first consciousness is perceptual; he becomes aware of what is before him, and that mental state kind gets correlated with whatever neural state kind happens to be going on in him at the time. Thereafter if (like Penfield) one artificially recreates that neural state kind one also brings on the mental event with which it has been correlated. This view of the matter might be called the hypothesis of Progressive Accidental Matching. It might seem to be an embarrassment for this view that there is as much as there is which is universally true of gross correlation: why is it for example that in all men it is activity in the occipital lobe that is correlated with visual experience? This embarrassment is easily overcome by the observation that the grosser neural tracts take parallel courses in all well-formed human brains: the optic nerves just do deliver their charges to the occipital lobe. Thus when a man sees for the first time it is inevitable that there is electrical activity in the occipital lobe. Thus the grosser sorts of correlation are universal because the gross structure of the brain is universal; it is the finer-grained correlations that are established by Progressive Accidental Matching.[18]

Let me now imagine an experiment to decide between these two competing hypotheses. Take a man who has been totally blind from birth through some retinal malfunction which can be corrected but hasn't been. Perform on him the Penfield experiment: electrically stimulate cells on his occipital lobe while he is conscious, and ask him to remember carefully the experience that stimulation occasions him. Then correct the retinal malfunction so that he can see; and in the early days after the restoration of his sight (before he goes mad) ask him whether this new experience of vision has anything in common with the experience he had, if any, when the Penfield experiment was done on him. If he says yes, then that will be a good degree of confirmation for the Preordination hypothesis.[19] If he says no, then that will perhaps be some confirmation for the Progressive Accidental Matching hypothesis: no mental event kind *is* the correlate of a neural event kind unless it *has become* the correlate of that neural event kind.

54

In fact, the experiment as I have described it is a somewhat inadequate test. For the neurons in the occipital lobe of a man blind from birth will have developed nothing like the complexity of interconnections (dendrites and axon-trees) which characterizes those of a man who has always seen.[20] None the less it is not a wholly inadequate test: for even an undeveloped occipital lobe is capable of mediating visual experience, for it does so in a baby when it sees for the first time. That is, we shall not expect our subject to seem to see trees or locusts or Amharic script under the stimulus of the Penfield electrode: we shall expect his visual experience to be as 'undifferentiated' as a baby's is said to be.

(All my intuitions – but they are only intuitions – tell me that this experiment will favour the Progressive Accidental Matching hypothesis; but the only neurophysiologist that I have been able to press on this issue ended by adopting – also by intuition – the Universal Preordination hypothesis.)

Now I said that these considerations about how correlations are established would lead us to glimpse a way of refuting the neurophysiological determinist. This refutation works by disarming the Correlation thesis, leaving the thesis of determinism in neurophysiology intact. The refutation depends upon two conditions: (a) that the Progressive Accidental Matching hypothesis is correct, and (b) that the style of correlation is generally non-additive, so that there is anyway not much correlation of higher mental species with higher neural species.

We saw in the first chapter that that step in the neurophysiological determinist's argument which embodies the correlation thesis has the form of strict implication: it says that it is *necessary* that if the neural state is such and such, then the mental state is such and such. Now if we adopt the Progressive Accidental Matching hypothesis, a correlation is not necessary the first time it is instantiated in a given person; it is necessary only on subsequent occasions; only after it has become 'established'. And if we hold that the style of correlation is non-additive, restricting ourselves pretty much to the basic Correlation thesis, then the only mental kinds that have correlates are total mental state kinds – *infimae species* – and these will very rarely be instantiated more than once in an ordinary human life: nearly all instantiations of total mental state kinds will be first instantiations. Thus very, very few logical connections between neural state kinds and mental state kinds will be connections of strict entailment: mostly they will be just conjunction.

So that, on these two conditions the neurophysiological determinist cannot get the necessary connection he requires between neural state kind and mental state kind, except in a very few cases.

We have seen that it is arguable that both these conditions hold (though it is also arguable that they do not), and this may therefore seem a satisfactory way of undoing the neurophysiological determinist, especially as it obviates the perhaps uncomfortable need to deny determinism in the physical universe: one can just claim that though the neural state must be what it is at a given time, the mental state is not bound thereby to be what *it* is at that time, at least not usually.

This would be a neat and agreeable way of making a case for freedom of decision, and it is a case for freedom of decision that we are out to make. But this solution will founder as soon as the libertarian, having thus made his case for freedom of decision, tries also to make a case for freedom of action. For the libertarian will not have obviated *physiological* determinism: a man's muscular activities will still be necessary occurrences.[21] Our libertarian will find himself in the same boat with the Stoics, holding that the acts of our bodies are determined but the decisions of our minds are free. And this is an awkward view; it seems flatly to contradict our ordinary experience,[22] and it can be saved from doing so only with some intricate theory that our acts are made the acts they are by our state of mind as we perform them, and that it is always true of any set of bodily motions that they can constitute several different acts depending what the state of mind behind them is. Such a theory as this is unlikely to be able to be made good, but even if it were made good, it would offer a degree of freedom much more restricted than the degree which we believe ourselves to possess. For we generally believe that we are free not only in the acts we perform but also in the bodily movements by which we perform them: 'I could have signalled with both hands'.

So that this method of making a case for freedom of decision by disarming the Correlation thesis seems pretty much a dead end for the libertarian.

(vii) The Identity theory

We have conducted the discussion so far without being committed – or without intending to be committed – to any view about the nature of mentalia or the nature of their relation (correlation is not a relation) to cerebralia. In this section I wish to adopt a non-reductive version of the

Identity theory. I do this because it provides the determinist with a more or less trouble-free account of how brain states produce mind states, obviating all the old difficulties about causal interaction. It also, surprisingly, provides the libertarian with a way of solving his biggest problem, as we shall see in chapter V. Having thus adopted the Identity theory I shall defend it against one of the usual and as yet not satisfactorily answered criticisms of it, namely, that there is no sense, or anyway no truth, to the notion that brain and mind states occur in the same place.

The non-reductive Identity theory says that a mental event and its correlated neural event are the same event under two different descriptions, a mental description and a neural description. These descriptions appear to have no logical relation to each other;[23] this is another way of saying that it is mysterious that such and such a mental event kind should be the correlate of such and such a neural event kind. The gravamen of the claim that these events are identical is a matter of some contention, and we must digress to examine it.

One's natural inclination is to suppose that identity for events works in the same way as it does for material objects, namely, that allegedly-two events are in fact identical if and only if they occur in the same place at the same time. So it was that the earlier identity theorists understood the matter, and they struggled in various ways to make good the claim that mental events and their neural correlates occur in the same place, and assumed that they occur at the same time. Davidson, however, has questioned this criterion for event identity and suggested another one.[24]

He has questioned it by offering this counter-example: if a ball is heated through 10C ° and at the same time turned 180° on one of its axes, here are two events which occur in the same place at the same time, and yet we should hardly want to call them one and the same event. Davidson avows incomplete confidence in the example for the reason that perhaps there *is* only one event here, the resultant change in the motions of the molecules of the ball. We can, however, find an even more convincing example. Suppose that a woman has a nervous disorder in her facial muscles which is such that whenever she blushes her face twitches: the extra blood sets off the nerves. 'She blushed' and 'her face twitched' describe two distinct events. And yet they occur in the same place, her face, at the same time. Here there is no temptation, I think, to see one event.

What Davidson suggests is that the criterion for identity of events has rather to do with sameness of causes and effects. Two events are identical if and only if anything which is a cause of the one is a cause also of the other, and anything which is an effect of the one is an effect also of the other:

$x = y$ if and only if $(\forall z)$ (z is a cause of $x \rightarrow z$ is a cause of y) and $(\forall z)$ (x causes $z \rightarrow y$ causes z).[25]

This criterion will certainly preclude the case of the ball from being a case of event identity. Will it also exclude the case of the blush and the twitch? One can set up the physiology of the example so that both events have the same set of causes (unless one says that, for example, muscle contraction is the cause of twitching); can one catch it out over the results? Do the blush and the twitch have the same results? One can I think argue that they do not, for the blush raises the temperature of the air, but the twitch does not; and the twitch sets up small shock waves, but the blush does not. Of course this claim to a difference in the results is a candidate for just the same reservations that Davidson voiced over his example of the ball. But surely other differences of result are likely; the twitch causes a crumb to fall off the woman's chin, for example, but the blush does not.

But Davidson's criterion is too strong. Thunder and lightning are archetypes of identical events, but they have very different effects. One cannot say that the lightning caused the blind man to stop his ears or that the thunder caused the corn to ripen,[26] or set fire to the barn it struck. This example seems to me conclusively to show that sameness of causes and effects is not a *necessary* condition of event identity, though doubtless it is a sufficient one. It does, however, suggest a new account of the necessary and sufficient conditions for event identity.

Why do we regard thunder and lightning as identical events? Surely it is because they have all the same causes, if not the same effects, and because they are spatio-temporally coincident. I suggest that it is these two conditions that are each necessary and that are jointly sufficient for event identity.

It seems to be true that spatio-temporal coincidence is a necessary condition of event identity. Davidson suggests, and retracts, a counter-example: I kill some astronauts by pouring poison into their spaceship's reserve water tank before it leaves earth. 'Pouring the poison' and 'killing the astronauts' are identical events, intuition tells us; but as the astronauts do not even begin to die for several months, say, are these acts really simultaneous? Yes, for it can be true that I have killed you before you die. To take a less extreme case: I stab someone and he goes to hospital in critical condition. The policeman who interviews me says, 'it may be that you have killed him', not 'it may be that you are killing him'. We do not know until he dies that I have killed him; but if he dies, then my act of killing him was accomplished well before he died. Ingenuity may furnish other counter-examples, but I can think of none.

And I can think of no case of event identity where there is even a temptation to say that any cause of the one is not also a cause of the other.

These two conditions may, then, be each necessary for event identity, but are they jointly sufficient? Our example of the twitch and the blush rather suggests not. For there we set up the physiology so that all causes were held in common. But can that really be done? Suppose that it is the blood's oxygen coming into contact with the neurons that causes them to fire and the muscle to contract. One cannot say that that chemical reaction involving oxygen is a cause of the blushing, can one? It seems that with a little thought we can find something which is a cause of the twitching which is not also a cause of the blushing. And the chances seem good also for showing that the coincidence criterion does not hold either. For the event of blushing can be said (I believe) to occur only in the outer layers of tissue on the face; the twitching however can be said to occur there and in the deeper muscle tissues. The case of the twitch and the blush seems not, then, to be a counter-example.

Tentatively, then, we can say that where x, y, z range over events:

$$x = y \text{ if and only if } (\forall z)(z \text{ caused } y \longleftrightarrow z \text{ caused } x),$$
$$\text{and } x \,\&\, y \text{ are spatio-temporally coincident.}$$

Now, what effect has the foregoing criterion for event identity on our contention that mental events are identical with their neural counterparts? The effect is curious. Let us assume for a moment that mental and correlative neural events are spatio-temporally coincident; I shall be defending this assumption shortly. Do a mental and its correlated neural event have all their causes in common? The answer seems to be that if the identity theory is true they do, and if it is false they do not.

The point is this. We do not know whether mental and correlated neural events are identical; it seems just as possible that they should be as that they should not be. We hunted over the terrain of event identity to find a criterion for it, and found one that seems satisfactory. But when we try to apply it to the case in hand we find we are just as uncertain whether the criterion is met as we are whether there is identity.

The criterion is adequate, but it does not in this case provide a serviceable test. (This is confirmation of the criterion's adequacy.) This seems like a case for the employment of a pragmatic principle: if the identity theory is a fruitful hypothesis, use it. Shortly, and again in chapter V, I shall show that it is a very fruitful hypothesis.

But first I must defend the assumption that mind states and their neural correlates coincide in space. Some have said that this claim is false, and some that it is even meaningless. I believe a good case can be made for saying that it is true. Let me, to ease the English, introduce the terms *homotopy* and *homotopic* to mean 'spatial coincidence' and 'spatially coincident' respectively.

The thesis of homotopy of mental and correlated neural events is held to be impugned by the fact that some mental events clearly do not occur in the same place as their neural counterparts. An instance is a pain in the toe. Presumably the neural activity corresponding to having a pain in the toe is much more widespread than just the nerves of the toe; indeed it is known to be so. The neural activity which is the correlate of the pain occurs also in the spinal cord and in the brain. Therefore the pain (mental event) and its neural correlate do not occur in the same place. An even starker example is that of a pain in a phantom limb. The neural activity which corresponds to a pain in the wrist of an amputated arm occurs in the central nervous system, and yet the pain occurs somewhere out in the air, in a place where no part of the body is. The mental event is not homotopic with its neural correlate.

The first of these examples provides a persuasive argument; the second makes the argument decidedly suspicious – are we really prepared to think that a pain in a phantom limb is a mental event which occurs out in the air? And reflection on the oddness of the latter example will show what is wrong with the whole argument.

The mental event is not the pain, but *having the pain*. In general, a mental event is not the content of an experience but the experiencing of that content. (This observation has, indeed, been made before.) If I see the Radcliffe Camera from All Souls, I have no inclination to say that that mental event occurs at the Radcliffe Camera; rather it occurs in me. The place of a mental event is not necessarily the place of its content.[27]

Now although a pain may occur in my toe, this does not imply that my having the pain occurs in my toe. If I have a watch in my pocket, the watch is in my pocket, but it would be odd to say that my state of having a watch occurs in my pocket.

Where, then, does the having the pain occur? Professor Shaffer claims that we have no conventional answer to this question, that we do not know what to say, and that we are therefore at liberty to establish a convention.[28] And a convenient convention to establish is that a mental event occurs at the place of its neural correlate. But in fact we needn't be quite so high-handed.

It seems to me certainly true that our ordinary intuitions do not readily yield a precise answer to the question 'where does such and such a mental event occur?' But that does not mean that an answer, precise or not, cannot be teased out of them. And teasing an answer out of our intuitions is not, of course, the same thing as legislating an answer. I shall argue that in fact an answer lurks in our intuitions, but that it is an imprecise answer. Further, I shall argue that its imprecision is appropriate.

It is often said that mental events are not extended in space, or, what comes to the same thing, that they are not situated in space (since, presumably, it will not be argued that they occur at an un-extended point in space). If this is so, then all spatial predicates attached to them will make equal nonsense. In a similar way, for example, lectures are not coloured, and the attachment of any colour predicate to a lecture makes nonsense, whatever the colour predicate that is attached. To say that Professor Jenkins's lecture was mauve is as nonsensical as to say that his lecture was green or red or yellow. Again, place predicates cannot be attached to days, and therefore it is as nonsensical to say 'Tuesday is in the beetroot salad' as it is to say 'Wednesday is in the blanket chest'. Again, the range of predicates having to do with marital status cannot be used of velocities, and 'the speed of light is a bachelor' is as nonsensical as 'the velocity of sound is divorced'. And so forth.

However, the case of the ascription of spatial predicates to mental events does not work in this way. If I am in All Souls looking at the Radcliffe Camera, it is nonsensical to say that the mental event of my seeing (seeming to see) the Radcliffe Camera is occurring in Peking; it is nonsensical to say that the mental event is occurring at the Ashmolean; but is it nonsensical, or simply wrong, to say that it is occurring at the Radcliffe Camera? Is it, finally, nonsensical to say that it is occurring in All Souls? I submit that not all spatial predicates applied to mental events are equally nonsensical. Further, it seems not at all nonsensical to think that my mental events are events which occur in my body.

61

Indeed this last observation even sounds as though it may be right. It seems to founder, though, when people ask exactly where in my body my mental events occur. For this question seems very hard to answer. The impossibility of answering this question has led people to suppose that the idea that mental events had a spatial position must have been wrong after all. But why should they not conclude instead that although it is true that my mental events occur in my body, they have no more precise place than just 'in my body'. This would not mean (a) that they occur at every place in my body; and, more importantly, it does not mean (b) that they have a precise place in my body but I am unable to discover it. Their spatial position, within certain boundaries, just is indefinite.

If such an answer can be allowed we shall be able, in a way that is agreeable to our intuitions, to say that our mental events occur in our bodies, and thus that they occur in space, but we shall not have to face the unanswerable next question, 'exactly where in your body do your mental events occur?' By not having to face that question we shall avoid the temptation to throw out the whole notion that mental events have spatial location at all. But can such an answer be allowed? Can we make any sense of the notion of an actually indefinite spatial location?

I shall argue that we can, and indeed that we very often do – that a great many events which we might suppose to have spatial location have, in fact, indefinite spatial location. Given the fact that a great many events have this feature it ought not to be particularly surprising that mental events are among them. The mystery of the notion of actually indefinite spatial position will be relieved when we see the generality of its instantiation.

Let us begin with the archetypal philosophical event, the collision of two billiard balls. Where does this event occur? At the point at which they meet? Over the infinitely thin plane surface over which the two balls, because of their elasticity, ultimately touch one another? In the whole space occupied by the two balls at the first moment of contact? In the whole space occupied by the two balls at the moment of their greatest compression? It seems, indeed, very hard to decide among all of these. And why should we decide? Each of them seems to have some claim to be considered; to legislate that one of them is *the* place where the collision occurs is to deny the claim of the others. Is it not more respecting of our intuitions to say that the collision occurs at some indefinite place within a certain region of space that encloses all the candidates – say, the cylinder with hemispherical ends which would contain both balls at the moment of contact. (This of course will only do for a head-on collision; other cases will be more complex.) In fact, though, we are never so precise even as this about the region within which the collision took place. If asked where it took place we might say 'over there' and point vaguely; we might even put a finger on a spot on the table and say 'here' - but a spot on the surface of the table manifestly is not where the collision took place. We are, I think, very willing to allow that the collision has spatial position; but it proves impossible, except by legislation, to say exactly what the spatial position is. We should hold greatest faith with our intuitions if we said that the spatial position of this event is, simply, indefinite. Similar arguments apply to a plethora of other cases: turning a page, lighting a pipe, eating a biscuit.

Those are cases where we are led to say that the spatial position of the event is indefinite because there are a number of candidates among which it seems impossible to decide. Here now is another sort of case. Suppose that a play is being performed. One might ask where, and have the answer 'in the Playhouse'. One might then go on to ask - as we ask about mental events - 'yes, but where in the Playhouse?' Already the question is a bit odd, but it might get the intelligible reply, 'on stage, of course'. The further question, 'yes, but where on stage?' would seem unanswerable. Not unanswerable because it is too hard to find out exactly where on stage characters stood or moved or where props and furniture were placed, but unanswerable because it is not clear that if there were, say, three little areas of the stage where no character or property ever stood or moved, it would be right to say that the performance of the play took place on the stage except for three little areas of it. The non-occupation of those areas could well have subserved the total dramatic effect. Or even if it didn't, why should they be excluded from being part of the place where the play was performed? I submit that on our intuitions they would not be excluded. The place where the play is performed is the stage; beyond that its position is simply indefinite.

Again, if a service takes place in a chapel, it is wrong to ask exactly where in the chapel it takes place. In a sense it takes place throughout the chapel - though that would be a strange thing to say, and it wouldn't imply that some person or thing which had to do with the service filled each lot of space in the chapel. Beyond just 'in the chapel' the position of the event, the service, is indefinite. This sort of example, too, can be multiplied.

A good many perfectly ordinary events, then, which we should all acknowledge to be events occurring in space, occur at actually indefinite spatial positions. Once their spatial positions are narrowed down to certain boundaries, to ask for further precision is to ask an unanswerable question, or at least one which is not answerable in perfect accord with our intuitions: to give an answer, an artificial selection from among a number of possibilities would have to be made. In these cases we do not, when faced with this unanswerable question, simply throw out the whole idea that these events are in space at all; we rather say that the question is not apt. Why then should we be terrorized by those who ask 'where precisely in your body do your mental events occur?' The question is not apt.

Now it may be thought that in defending in this way the view that mental events have spatial position I have undermined the case for their being homotopic with neural correlates. For if homotopy is to be claimed and a mental event occurs in indefinite spatial position, must not the correlated neural event also occur in indefinite spatial position? Is there any case for thinking that neural correlates occur in indefinite spatial position?

Indeed there is, as we have seen in our discussions of how the one-to-one correlation of neural and mental event kinds could work out. We saw there that to the extent that the style of the correlation is additive, the OFF neurons count in a sense and do not count in a sense. The case is analogous to that of the untrodden corner of the stage in a play: the performance does not occupy it in the same way that it occupies the rest of the stage, but the performance will still be said to occur 'on the stage' – not 'on the stage except for one corner'.

I conclude that the spatial coincidence condition for mind–brain identity is satisfied. Mental events occur in space, but in indefinite spatial position. The indefiniteness of their spatial position is a feature they share with many other events. Neural correlates also occur in indefinite spatial position; and the two indefinite spaces are enclosed by the same firm boundary, the body. Mental events are homotopic with their neural correlates.[29]

I said that after defending the claim that mental and correlated neural events are spatially coincident I would show that the Identity theory is a fruitful hypothesis, and that since none of its necessary conditions obviously fails to be satisfied, it should be assumed. The way in which it is fruitful is this: it replaces something which can seem conceptually impossible with something that is merely mysterious.

Any theory of mind has to come to terms with interactionism, the phenomenon. Bodily changes just do seem to bring about mental changes, and mental changes just do seem to bring about bodily changes. One way of coming to terms with interactionism the phenomenon is to advocate Interactionism the theory: that mind and body have straightforward causal relationships. Now this theory is capable of seeming so impossible of truth that it drove Geulincx and Malebranche, for example, to their metaphysically lumbering Occasionalism, the view that God effects each causal link between the two. Now I do not want here to claim that Interactionism is impossible;[30] but certainly it can strongly seem so. What I can securely claim is that it is a big philosophical trouble spot. The advantage of the Identity theory is that it completely obviates the philosophical trouble spot, and its only residue of embarrassment is the mystery as to why the correlations are what they are. Mysteries can be lived with - we live with the mystery that the frequency of yellow light is 5.2×10^{14} cycles per second - but conceptual impossibility, or even the appearance of it, cannot be endured. The Identity theory, then, in a way that is well known, accommodates interactionism the phenomenon with ease. A mind event at t causes a brain event at t' just because the mind event *is* a brain event at t, and there is no mystery about a brain event at t causing a brain event at t'. Because the Identity theory affords such an easy solution to this intractable problem, the neurophysiological determinist ought to assume it.

(viii) Summary

We have now looked at the Correlation thesis, and its metaphysical adjunct the Identity theory, in some detail. They seem to make a strong case for the neurophysiological determinist, if taken together with the thesis of determinism in neurophysiology. The only proviso is that the Correlation thesis must be strong enough to ensure that the correlation of particular mental and neural events is necessary. This requires either (a) that the Preordination hypothesis be true, or (b) if it is rather the Progressive Accidental Matching hypothesis that is true, that there be correlation of superior mental and neural species; this last requires in turn that the style of the correlation be to some degree additive.[31]

But if the Correlation thesis is true in some such strong version it constitutes more of a threat to the libertarian than just that threat which is constituted by its being one of the supports of his opponent's view. For if the Correlation thesis is true,[32] then there seems no chance for the libertarian claim that our decisions are free: if determinism in neurophysiology is true our decisions will be causally determined; if it is false our decisions will be random. And randomness is not freedom. The thesis of determinism in neurophysiology must, however, be either true or false. Therefore the Correlation thesis entails that freedom of decision is impossible – causal determination or randomness seem to be the only alternatives.

We shall for what follows assume that the Correlation thesis and the Identity theory are true; and we shall find a way for the libertarian to break out of the dilemma in which he is caught.

IV

THREE PROBLEMS FOR THE LIBERTARIAN

(i) Neurophysiological indeterminism

The first task before the libertarian is to break the neurophysiological determinist's argument to show, of any decision, that it had necessarily to occur. We saw in the last chapter that it would be short-sighted of him to do this by denying the Correlation thesis, for then he would still have physiological determinism to contend with, and that contention would not be easy. The only other breakable thing in the neurophysiological determinist's apparatus is the thesis of determinism in neurophysiology. The libertarian theory of free decision which I advocate breaks the determinist view at this latter point, and leaves the Correlation thesis intact.

The libertarian, that is, will claim that not all neural events or states are completely determined by preceding physical states, that there is an element in the nervous system of real indeterminacy. About this claim several things need to be said.

First of all, the indeterminacy in question is real, not epistemic: it is indeterminacy and not uncertainty. In the language which the Identity theory causes us to adopt, this claim would be: there are some states of the nervous system n, such that there can be in the physical language no descriptions of them and of any physical states $1 \ldots n-1$, such that the conditional sentence whose protases mention the occurrence of those previous $n-1$ states under such descriptions, and whose apodosis mentions the occurrence of the nth state under such a description is, or states, a law of nature.

The stipulation that there *can* be no descriptions is designed to rule out the reflex: 'well, of course these events aren't uncaused, it's just that we don't know what causes them'.

Second, the indeterminacy in question is what is sometimes called partial indeterminacy. That is, it is not the case that at a certain moment in the nervous system literally anything may happen next – rather that several different things could happen next and one of them will.[1] The form of the causal law governing the events of some intervals in the history of the nervous system is not p $\rightarrow$ q, but p $\rightarrow$ (qvr). Occasionally one hears the libertarian view derided on the grounds that it would entail a kind of spastic, disconnected mental or even muscular life; but this second observation should remove any concern that we might be in for that kind of result.

Third, the libertarian ought to offer some plausible account of just where and how in the nervous system these indeterminacies could occur. After all, the indeterminacies recognized by physics[2] occur in the motions of particles much much tinier than those in which neurophysiologists generally deal. I shall offer two such ideas.

The first is an enlargement upon a sketchy suggestion made by J. C. Eccles.[3] In the end, what makes for different states of the nervous system is whether or not, and when, a neuron fires. What determines whether or not it fires is whether or not it receives enough input of electrical stimulation from other firing cells; when the stimulation reaches a critical level the cell fires, communicating its charge to other cells with which it is in contact. Whether or not the critical level of stimulation is reached depends upon how many other cells deliver charges, and whether they deliver them within a sufficiently short interval of time. That is, suppose a cell needs ten charges to cause it to fire; it will have to have those charges delivered within an interval of, say, one-tenth of a second, otherwise the earlier charges will have leaked away by the time the later ones arrive, and the required ten will not be built up.

The principle is rather like that of a self-emptying tank which leaks: unless water is delivered to it at more than the rate of the leak the tank will never be full enough to trigger the self-emptying mechanism. We can press the analogy by supposing that the water is delivered to the tank not in a continuous stream, but by cupfuls. Clearly, the tank's self-emptying device will not be set off unless enough cupfuls in a sufficiently short space of time, are delivered to it.

Now suppose we work out a way in which it can be undetermined not whether, but just when, a cell will deliver its charge to another.

Suppose that other cell has nine of the ten charges that are required to make it fire; clearly the indeterminacy as to when the tenth charge is delivered produces in this case an indeterminacy as to whether the cell in question fires that time round. For if the charge is delivered sooner the cell will fire; and if it is delivered later, it will not. Following Eccles, we can describe the neuron that has only nine of its required ten charges and needs the other in short order, as 'critically poised'.

Now the indeterminacy as to when the charge is delivered can in turn be derived from an indeterminacy of position, an indeterminacy where.

The transmission of charge from one neuron to the next is not accomplished by direct contact: there is a space between neurons, called the 'synaptic cleft', at the place where charges are transmitted. Into this cleft the sending neuron releases a small amount of a chemical transmitter substance; the transmitter crosses the cleft to the receiving neuron and there induces charge. The chemical transmitter in the sending neuron is packaged in small envelopes, 'synaptic vesicles', which are situated at or near the membrane wall between the sending neuron and the synaptic cleft; when stimulated the vesicle moves to the membrane wall and releases its chemical across it. This process is depicted in Figure 4.1.[4]

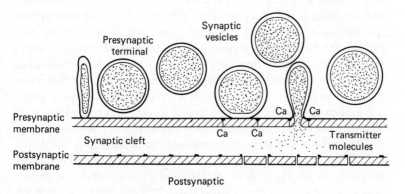

Figure 4.1 The transmission of charge

Eccles calculates that there is an indeterminacy in the position of a synaptic vesicle of about fifty ångström in one millisecond; now if the position of a vesicle is indeterminate, it is also indeterminate *when* it will reach the presynaptic membrane to release its chemical. Thus the indeterminacy as to when the charge is delivered can be derived from the indeterminacy where the vesicle is located.

My second suggestion of how the needed indeterminacy in the nervous system might occur is a bit simpler. It is that it could be the case that the critical level of excitation for making a neuron fire is not exactly a critical level, but a threshold. The situation might be this: a given neuron may fire after 97 units of charge are built up in it, and it must fire once 103 or more units of charge are built up in it, but for 97–102 units it may or may not fire: whether it fires or not is quite simply undetermined. We shall look further at this suggestion in a moment.

It might seem that all this is inadequate: that the firing or not firing of one neuron will make little difference to the total neural – hence to the total mental – state. On the contrary, Eccles has demonstrated that such a small thing can have an appreciable effect over a sizeable part of the brain very quickly indeed.[5] And in any case more than one neuron may be 'critically poised' at any one time.

The fourth thing needing to be said about the libertarian's claim that there is indeterminacy in the nervous system is an anticipation of an objection. I am not sure that I have ever seen this objection stated explicitly in anti-libertarian writings, but it is something which always used to make libertarianism, to me at any rate, seem wildly implausible. The objection is this. Suppose there does occur in the nervous system the sort of undetermined event we have postulated; suppose that periodically the nervous system takes an unpredictable leap – like a Mexican jumping bean. How does the mind arrange to have its decisions occurring simultaneously with that unpredictable leap, so that they, too, may be undetermined? Surely it is as unpredictable when the leap will happen as it is unpredictable what the leap will be. The libertarian seems here to be requiring an incredible degree of coincidence.

The reply is a matter of considerable importance for libertarian theory. Let us look again at my second suggestion how indeterminacy occurs in the nervous system. At first glance one supposes that what is undetermined is when the neurone fires – or if the charge fails to reach 103 units, having reached 97, whether it fires. One supposes that the undetermined is when the neuron fires – or if the charge fails to reach has reached 97 units, it is just as undetermined that the neuron is *not* firing, if it is not. That is, while the charge is between 97 and 103, the conditions are sufficient neither for it to fire nor for it not to fire. Whatever the neuron is doing for the interval from 97 to 103 units of charge is not physically explicable, not caused. So that it is not a Mexican-jumping-bean-style of undetermined event that is being postulated, but an enduring undetermined state. The same observation applies, *mutatis mutandis,* to my first suggestion as to how the neural indeterminacy might occur. If there are thus enduring undetermined states it is much less a coincidence that decisions occur in them. In chapter VIII I shall, indeed, argue that it is not only at the moment of decisions that we have liberty of indifference, but more or less all the time.

But of course it is to be noted that we are not by this committed to thinking that all neurons are thus in undetermined states most of the time: that really would result in chaos. For of course it may be that generally neurons receive – to recur to the example – rather more than the 103 units of charge sufficient to make them fire, and that they receive them all at once, so that there is no interval of indeterminacy.

It is true that all this is highly speculative: how could it be otherwise? And of course the libertarian must not make himself ridiculous by clinging too desperately to these speculations. But it seems reasonable to expect him to say something plausible about his postulation of indeterminism in the nervous system.

(ii) The mind's dependence on the brain

The second problem facing the libertarian is that of analysing and getting around the standard objection that we are no nearer freedom with *random* neural events than we were with causally determined ones: if our decisions are what they are because by pure chance a neuron fired sooner rather than later, then the responsibility for that decision is to be laid at the feet of that undetermined physical event: we are delivered out of the bondage of causal determinism only to find ourselves in the bondage of randomness. This objection seems to me powerful but obscure; it must be analysed before it can be answered.

What seems to lie behind it is the idea that mental events are *produced* by neural ones, that mental ones are to be explained by reference to neural ones, that mental ones are what they are because the correlated neural ones are what they are, and not vice-versa. This notion loses some of its power as soon as we put it into the language of the Identity theory: 'A certain mental description is instantiated *because* its correlated neural description is instantiated, and not vice-versa.' However that may be, the general notion behind the objection is that the mind is brain-dependent, but that the brain is not mind-dependent: one isn't happy to say that the neuron fired when it did because the decision was what it was; but one is happy to say that the decision was what it was because the neuron fired when it did, even when its firing when it did was not physically determined. Why should this be?

The next chapter will be given to some intricate negotiations of this matter, ones which will eventually yield an answer to this ancient and central objection to libertarianism.[6] In the meantime I wish to examine two famous attempts to show that the mind is not dependent on the brain, at any rate, not utterly so. The first is J. R. Lucas's Gödelian argument; the second, D. M. MacKay's idea of 'relativistic logic'.

Lucas's Gödelian argument

Lucas argues[7] that a mind must be more than a machine, and thus more than its own machinery, because a mind-machine is subject to Gödelian incompleteness, but a mind is not. That is, there are certain statements which can be believed by a mind but which a symbolic system, a logistic calculus, cannot prove, or what comes to the same thing, generate. If we set up such a system L, which so operates as to infer in the way that a given human mind infers (whether rightly or wrongly by external standards) it will be found that there are some propositions which can be believed to be true by the human mind, but which cannot be proved in the logistic calculus L. Thus the human mind can infer a proposition which its parallel machine cannot generate.

The proposition in question is 'this formula cannot be proved in the logistic calculus L', a self-referent proposition. Lucas claims that this proposition can be believed by a human mind, can be seen to be true, but that a logistic calculus L, by Gödel's theorem, cannot generate it.

Now I shall exhibit and criticize the more detailed argument in a moment, but first of all let me comment that if Lucas's argument is right, it proves only that in this one rather rarefied case the mind is not tied by its machinery, but can, so to speak, rise above it; it proves nothing about the other and very much more common mental activities.[8] Still, if the mind can be shown conclusively to be sometimes not bound by its machinery, that is at least evidence that in general it is not so bound.

I don't think, however, that Lucas's argument does show at all that the mind is sometimes independent of its machinery. Now of course many have held this view, and the issue has been much discussed in the literature; but none of those who have offered refutations of Lucas's argument has said what is really wrong with it, but only shown that it must be wrong. I have a suggestion as to what is wrong: briefly, the part about the inadequacy of the logistic calculus L is entirely correct, but this little inadequacy of L feeds on a much more general inadequacy – and thus Lucas's case is so far stronger than he realizes; however L can be reformed in a very simple way which will remove the general and therefore also the particular inadequacies, and this reformation of L will not alter its character of being machinery. I shall now explain and expound this more carefully.

Lucas's argument is this. We set up a logistic calculus L whose inferential operations reflect the inferential operations of a given human mind. Thus, *ex hypothesi*, if:

(1) the mind believes proposition p, then

(2) formula p is generated by (deduced in, proved in) L.

Now if we substitute for the general case of p the proposition 'this formula is unprovable in L', we have, if:

(3) the mind believes that 'this formula is unprovable in L', then

(4) 'this formula is unprovable in L' is provable in L.

Lucas argues that (3) can clearly be the case, for we ourselves (our minds) can see that the proposition 'this formula is unprovable in L' is true, but (4) could never be the case since it is contrary to Gödel's theorem. 'This formula is unprovable in L' is simply not provable in L; L is necessarily incomplete in this respect. Thus the sought-after calculus machine model of the mind inevitably is inadequate: the general scheme of translatability proposed in (1) and (2) simply cannot be carried through for all p: there is at least one belief of which the mind is capable but which the logistic calculus is unable to generate.

Let me now show more clearly how and why the inadequacy in L arises and how it can be overcome. Let us look again at (4), which should have been the L-translation of (3) but which L, being Gödelianly incomplete, could not yield.

(4) 'this formula is unprovable in L' is provable in L.

Let us retranslate (4) back into mind-language. The obvious rendering is (3), the mind believes that 'this formula is unprovable in L'; but a second rendering is also possible, namely,

(5) the mind believes that 'the mind disbelieves this proposition'.

Thus (4) is an ambiguous rendering, being a possible rendering of two different beliefs of the mind. Lucas's argument is grounded in the difference in truth value between (3) (true) and (4) (false). There is, however, no difference in truth value between (4) (false) and (5) (false):

with respect to (4) and (5), the alternative re-rendering of (4), there is no inadequacy in the logistic calculus. Moreover, this ambiguity of L-rendering is systematic: any intensional statement on the mind side whose embedded clause is in L-language will have the same L-equivalent as a similar intensional statement on the mind side whose embedded clause is put into mind language. Thus, for example, if the Gödel number 123456 is the equivalent of the proposition 'this cow is ill',

(6) 123456 is generable in L

on the L-side will be ambiguous, being a rendering of either

(7) 123456 is believed by the mind, or

(8) 'this cow is ill' is believed by the mind,

on the mind side. Both (6) and (8) could be true, while (7) could well be false, indeed must be false in any case where the mind in question is not fluent in L.

Here is an analogous case which may provide illumination:

(9) 'Ceci n'est pas une phrase française' is not an English sentence

would be rendered into French:

(10) 'Ceci n'est pas une phrase française' n'est pas une phrase française.

To translate the main clause of (9) 'is not an English sentence' by 'n'est pas une phrase française' is the same convention as that of translating 'the mind believes' by 'is provable in L'; and though the convention is startling at first, the aptness of it becomes clear on reflection. However, the French sentence (10) is capable of an English rendering other than (9), to wit:

(11) 'This is not an English sentence' is not an English sentence.

Here (9) is true, but both (10) and (11) are false; it looks as though English is capable of uttering a truth (9) which French, under these conventions of translation, cannot reproduce.

So far we have merely underlined Lucas's argument, though I think that the English–French analogy also undermines its power to persuade us that there is something essentially inadequate about logistic calculi as representations of the functioning of the mind: we are not tempted, unless by an outrageous sort of chauvinism, to suppose that there is something essentially or deeply inadequate about French.

Now I claimed above that Lucas's argument feeds on, because it is a particular example of, a more general inadequacy indigenous to the attempt to translate intensional statements of this sort under these conventions of translation. Let me now say how and why this is so. It should by now be clear that it is possible to find an infinite number of cases in which the statement on the Right side of a translation table is ambiguous because it has two possible different renderings on the Left side. Ambiguity is a kind of inadequacy, but not perhaps as sharp an inadequacy as the one which Lucas exhibits. The sharp inadequacy of Lucas's example rests on two extra conditions present in it. The first extra condition is the apportionment of truth values among the three statements (3), (4), and (5). Because the ambiguous rendering on the L-side is false and because one of the renderings on the mind side is true, the language of the mind side has a capacity to utter a truth which is lacking to the language of the L-side. Now in these cases of triplets bearing the relation that one is an ambiguous translation of the other two, all the imaginable permutations of truth values are, I think, possible. But it is *this* apportionment of truth values, wherein the ambiguous translation on the Right side is false but at least one of the statements on the Left side is true, that founds the sharper inadequacy of the language on the Right side than mere ambiguity. But there is a second feature of the example which Lucas chose that makes the inadequacy seem sharper still and his argument the more persuasive. It is that the falsehood of (4) is a *necessary* falsehood. A contingent falsehood would have made the same point, but with more strain on the imagination and therefore with less dialectical force. Here is an example of such a triplet in which the falsehood of the L-side is merely contingent:

(12) the mind believes that 123456   (T)

(13) 123456 is generable in L   (F)

(14) the mind believes that this cow is ill   (F)

Nor does this triplet contradict the whole hypothesis about what L can do: the foundation of the falsehood of (13) may be the falsehood of (14); and it may be that the L-equivalent of (12) is not (13) but:

(15) 789101112 is generable in L.

However, the introduction of (15) to explain the falsehood of (13) when (12) is true seems gravely to undermine Lucas's argument, for it suggests the need for the introduction of a distinction between the sense and the reference of statements, on the lack of which distinction the argument can be said to rest.

I shall not pursue the argument in that direction, but rather I wish to make a certain alteration in L which will avoid the Gödelian difficulty in Lucas's example. Let me say first, however, that I make this alteration on the quite general grounds that there is a systematic ambiguity built into the translation convention under which we have been operating. It happens that by legislation which removes this general ambiguity we can also remove the Gödelian inadequacy. I emphasize that this legislation is advanced on quite general grounds.

The reform which I propose for L is simply its duplication. Let L be the collective name for two logistic calculi, L_1 and L_2. L_1 and L_2 are identical, except that L_1 can generate the truths about L_2 which L_2 cannot generate about itself, and vice-versa. The combination of L_1 and L_2 will be the machine which can infer all that the mind infers.

Let us work this out for the English–French analogy first of all. We say that we have two languages, English and Substitute; Substitute consists, say, of French and German, rather as reformed L consists of L_1 and L_2. In the analogy

(9) 'Ceci n'est pas une phrase française' is not an English sentence
and
(11) 'This is not an English sentence' is not an English sentence
had both to be rendered into French by

(10) 'Ceci n'est pas une phrase française' n'est pas une phrase française,

which, therefore, was ambiguous.

However, if we work the translations in Substitute, (9) becomes:

(16) 'Ceci n'est pas une phrase française' ist kein deutscher Satz
and (11) becomes:

(17) 'Dies ist kein deutscher Satz' ist kein deutscher Satz
or:

(10) 'Ceci n'est pas une phrase française' n'est pas une phrase française.

And it is not the case that (11) is ambiguous as an English translation since it translates either (17) or (10) in Substitute, for (17) and (10) are synonymous and in the same language, namely, Substitute.

Let me illustrate this last point with another example. Although it is true that on a French/English translation table, 'blind' on the English side is ambiguous because it does for both 'aveugle' and 'persienne' on the French side; it is not true that 'moose' on the English side is ambiguous because it does for 'élan' and 'orignac' on the French side: 'élan' and 'orignac' are synonymous. So, too, (17) and (10) are synonymous statements in Substitute.

Thus a language like Substitute, being a conjunction of two languages each of which can talk about the other, relieves the systematic ambiguity of translation in general and the Gödelian inadequacy in particular.

Have we, however, here suggested something which entails an infinite regress? If so, it can, I believe, be escaped with a device.

It might be argued that a language like Substitute will deal with the ambiguity problems for sentences like (9) ' "Ceci n'est pas une phrase française" is not an English sentence' by putting it all into the other wing of Substitute, namely, German; but will Substitute be able to deal with, for example:

(18) ' "Ceci n'est pas une phrase française" ist kein deutscher Satz' is not an English sentence?

Would not a Substitute which could unambiguously translate this sentence have to have three sub-languages in it? And is it not possible to carry on making many-layered intensional sentences for translation, so that the number of sub-languages in Substitute would ultimately have to be infinite?

This infinite regress can be escaped. (18) can be accommodated in the present two-winged Substitute as a conjunction of two statements in that language:

(19) (a) ' "Ceci n'est pas une phrase française" ist kein deutscher Satz' ist kein deutscher Satz,

and

(b) ' "Ceci n'est pas une phrase française" ist kein deutscher Satz' n'est pas une phrase française.

The rule of this translation is that in (a) whatever was in English in the original is put into German, and in (b) whatever was in English in the original is put into French. Clauses which in the original were in one of the sub-languages of Substitute are left in that sub-language in both (a) and (b). Thus if a clause in (a) is in a different sub-language from its parallel clause in (b), that clause was in English in the original; if parallel clauses are in the same sub-language in both (a) and (b), then the parallel clause in the original was in that sub-language. Short reflection will show that this technique will accommodate all the possible permutations of the three languages in a two-tier (three-clause) intensional statement: a unique translation in Substitute is available for every permutation. And Substitute can similarly accommodate intensional statements of more tiers than two.

Therefore, I conclude that the general ambiguity built into the translation procedure advocated by Lucas can be overcome by this duplication of L; and once the general ambiguity is thus overcome, the Gödelian inadequacy is also overcome. The mind statement

(3) The mind believes that this formula is unprovable in L

which was alleged to be untranslatable into L, is in fact translatable into reformed L:

(20) 'this formula is unprovable in L_1' is provable in L_2

There follows as a corollary of all this a much more succinct refutation of Lucas's position. Lucas set up, as it were, a translation table with a mind on one side and a machine on the other. The machine was found to be inadequate to reproduce all that the mind could produce by its inferential operations. Therefore, Lucas concluded that the mind, since it has a certain extra capacity, is not like a machine. However if the table had been set up with two machines, one on each side, and no mind present, the same inadequacy would have been present in the translation of the operations of the one into the other. The conclusion is that the one machine is greater than the other, since it has a certain extra capacity. Therefore, the mind's property of being greater than a machine is one which it shares with other machines. Therefore the mind is not greater than any machine whatever.[9]

MacKay's 'relativistic logic'

The second famous argument to show that the mind is not dependent on the brain is that of D. M. MacKay, which introduces the idea of 'relativistic logic'.[10]

MacKay argues that the individual is free from neurophysiological determinism by claiming that a man can never know the state of his own central nervous system because by so knowing he alters it and therefore does not know it as it is – does not, in fact, 'know' it at all; if he then tries to know this altered state of his CNS, he alters *it,* and so *ad infinitum.* Since he can never know the state of his own CNS he cannot neurologically predict his own behaviour: he is not and cannot be in possession of the information necessary to make that prediction. A person could, however, know the state of the CNS of another, and accurately predict his behaviour on the basis of that knowledge. Such a statement of prediction would then be false if the agent believed it, but true if only another person believed it. This escape from the clutches of neurophysiological determinism lands us thus in the need to accept a relativistic logic whereby the truth or falsity of a certain proposition depends upon who it is that is entertaining that proposition.

The argument assumes, I think, that it is important for the task of prediction that the predictor (agent or observer) know the state of the CNS as it is at the time he knows it, and not as it was, say, a millisecond ago. For presumably there is no difficulty in an agent's believing a description of the immediately past state of his CNS. The argument then holds that there is a reason of logic why an agent cannot know the state of his own CNS at the moment he is knowing it; we can follow the introduction of the regress easily enough. This reason of logic does not, however, arise for a second party: X's coming to know Y's neural state does not *eo ipso* work a change in Y's neural state.

Now I think this argument is wrong on two scores. The first is that there is no logical difficulty about knowing the state of one's own CNS as it is at the time of being known; the second is that the initial assumption - that knowledge of the CNS state as it is at the time of being known is necessary for prediction - is ill-founded.

It is logically possible simultaneously to have and to know that one has a given total neural state. Let there be a neural state *j* which is the correlate of this mental state: knowing that one is in neural state *j*. Here one couldn't have neural state *j* unless one knew one had neural state *j*. There seems to be nothing contradictory of itself in this idea. If one adheres very firmly to a causal condition for knowledge, i.e. that the known must be one of the causes of the knowledge of it, and if one believes that causes must precede their effects, then the notion of neural state *j* will be troublesome: one will have to revise the claim about it to this, that *j* is the correlate of *believing* that one is in neural state *j*. But it seems to me that our intuitions that the case of *j* is a case of knowledge pull stronger than intuitions about a causal theory of knowledge and the principle that causes must precede their effects. At any rate, it is clear that it is possible simultaneously to have neural state *j* and to believe truly that one has neural state *j*, and this, I think, is all that the refutation of MacKay here would require.

But why must the knowledge be simultaneous with the neural state known for prediction to be possible? Surely knowledge of an immediately past neural state is all that is needed, and that is equally available to agent and second party. If an instantaneous picture of the CNS which contains all the information necessary to prediction (i.e. presence, position, and strength of action potentials etc.) is fed to a computer, the computer can, as it were, simulate the immediate future of the CNS but at a considerably accelerated rate: charges travel much faster in wires than in neurons. Admittedly, this predication could not foresee new inputs into the CNS, but if we deal with sufficiently short stretches of time, that need not be significant. There is, of course, one significant kind of input whose effect the predicting computer could foresee, and that is the input to the CNS of the prediction information which it has just given to the agent.

Does this last observation give rise to an infinite regress? Let us work it out.

Suppose that at time t_1 I begin deliberating over a decision between doing x and doing y. At the same moment t_1 the state of my CNS is reproduced in a calculating simulator (in fact, say, $t_1 + 10^{-9}$ seconds). Except for the reports of the simulator no new information will be introduced to my CNS between t_1 and the making the decision. The simulator calculates that I shall decide x at t_3 and reports this to me at t_2. Immediately the simulator returns to t_1 and works through again, this time incorporating the effect in my CNS of hearing at t_2 the prediction that I would decide x at t_3. At t_3 it reports its altered prediction that I will decide y at t_4. Immediately it returns to t_1 and works through again, this time incorporating also the effect in my CNS of its report that I will decide y at t_4. And so forth. It looks as if the simulator is making valid predictions which are only invalidated by their being reported to me. This state of affairs has the same structure as the one which led MacKay to his relativistic logic. He was led there by the (incorrect) thought that a description of my neural state is falsified by my believing it to be true; are we to be led there by the thought that a prediction of my behaviour is falsified by my believing it to be true?

Let us investigate by imagining the simulator to be more dissembling. Let it report only a prediction which already takes into account the effects of its own being reported. Is this a possible mechanism?

There are several cases. The simulator might forecalculate the effects on me of which prediction it reports. This calculation would, in the case of a simple decision between x and y, have one of four possible results:

(1) if it predicts x, I do x
    if it predicts y, I do y
(2) if it predicts x, I do y
    if it predicts y, I do y
(3) if it predicts x, I do x
    if it predicts y, I do x
(4) if it predicts x, I do y
    if it predicts y, I do x.

Now clearly if the outcome of the simulator's calculation of the effect on my CNS of receiving a prediction is (1) or (2) or (3), the simulator can straightforwardly make a true prediction which already takes into account the effect on me of hearing that prediction. In case (1) this is done by predicting either x or y, in (2) by predicting y, in (3) by predicting x.

Therefore, in these three out of four paradigm cases it would be right for me to believe a prediction of my decision; I should not thereby falsify it. MacKay is thus wrong, after all, to hold that 'no complete prediction of the future state of the organizing system is deducible upon which both agent and observer could correctly agree.' In some cases, at least, such a prediction is possible.

Can MacKay's view, however, be refuted for all cases: what of case (4)? It seems impossible here for the simulator to make me a prediction which already takes account of the effect of its being heard by me, and which is true. For it knows that if it says x I will do y, and if it says y I will do x. Perhaps the simulator could take refuge in an oracular pronouncement like 'you will do the opposite of what I say you will do; you will do x', which can claim to be a true prediction no matter what happens; but that doesn't quite seem to be prediction in the meaning of the act. We have not shown, then, that MacKay's regress can never work, but we have shown that it cannot always work.

We must, then, grant that in some cases it is impossible for an agent's future actions to be predicted truly to him by calculation of the future of his nervous system. But what has this to do with relativistic logic?

MacKay does not clearly explain exactly what he intends by this term 'relativistic logic'. One can, however, glean that it is an instrument for allowing two contradictory propositions both to be true, though not in an absolute way; and the relativity is to the persons who believe them. A prediction of an agent's decision is not true for the agent to believe (we have seen that this is not unqualifiedly so); but it is, or may be, true for other persons to believe. A given proposition, then, would count as a false belief for one person and a true belief for all other persons. This seems to be the kind of logical situation that MacKay has in mind.

But of course this is not a fair description of the case. For it is not that a given prediction statement constitutes a false belief for one person (the agent) and a true belief for all others, but that if the agent hears of the statement it may thereon become false, not only for him, but *for everyone else as well* (absolutely false); but if the agent does not hear the statement then it remains true for everyone else (absolutely true). The statement is not false for one and true for others, but false for everyone if it is heard by the agent, and true for all the others if it is not heard by the agent. This, however, is not a case of anything that could aptly be called 'relativistic logic'; it would more aptly be called a case of a performative audition. A statement's being heard by a certain person changes its truth value. There are other examples of this sort of statement – for example, 'You do not hear me', or 'I am in your good books, you great baboon', etc. And we have no tendency to suppose that these latter statements require a relativistic logic. It is just that there are certain statements which have, among their truth conditions, that they not be heard by certain people.

In fairness to MacKay I must say that he acknowledges the point that is italicized in the previous paragraph. But that acknowledgment does not appear to change his mind about the relativistic logic that he seems bent upon introducing. That can only be by lack of clear-sightedness.

I want now to notice, quite idly, an interesting result which could have been got from such a relativistic logic, had its introduction been successful. With a relativistic logic, we should have to admit that the following two contradictory propositions are both qualifiedly true: (a) X's decision to do y was a free decision, and (b) X's decision to do y was causally determined. As far as X is concerned, (a) is true, and as far as everybody else is concerned (b) is true. The first clause of the last sentence does not mean that (a) is not true but X thinks wrongly that it is; rather it means that (a) is true relatively to X. There will be no un-qualified predicates 'true' or 'false' in this case, but only qualified ones, if the relativistic logic is allowed. Now, if it is the case that people are responsible for their free decisions, and that if they are responsible for a wicked decision they can justly be punished for it, then the following enthymemic sorites might be set up (the qualified truth of the premise implying the similarly qualified truth of the conclusion):

(1) X's decision to do y was free.   (true relatively to X)

∴ (2) X is responsible for his wicked decision to do y.   (true relatively to X)

∴ (3) X can justly be punished for his decision to do y.   (true relatively to X)

Now if it is true relatively to X that X can justly be punished for his decision to do y, then he is scarcely in a position to complain when that punishment is administered; rather he should acquiesce, indeed rejoice, in its justice. X will have no grounds for complaint. And he will have no grounds for complaint even when his punishment is unjust as far as everyone else is concerned, because it is false relatively to everyone else that his wicked decision was free.

Everyone else, of course, includes God. So that we have here a logical device which would permit omniscient God to hold that X's wicked decision was determined and yet to punish him for it without inconsistency. There would be no inconsistency because He would hold the first on one logic and do the second on the other (since He is omniscient, both are accessible to Him). Thus Predestination becomes compatible with suffering Eternal Deserts. There lurks in relativistic logic an elegant and sophisticated argument for the central tenet of Calvinism. We have not, however, seen any reason why such a relativistic logic should be introduced.

Neither Lucas's argument nor MacKay's is successful, then, in showing that the mind is somehow independent of the brain. In the next chapter I shall offer a theory according to which it will seem reasonable to say that, sometimes, the brain state is dependent on the mind state.

(iii) The random and the free

But even after the ascendancy of mind over brain has been thus negotiated, there will still remain a large problem for the libertarian. This problem has nothing to do with the relation between body and mind; it arises as much for the libertarian who is pleading against psychological determinism as for the libertarian we are defending here. It is the problem of the difference between randomness and freedom in the decision itself; there certainly seems to be no compulsion in most of our decisions; and we have denied that psychological determinism is true; is not a decision, then, merely a random, undetermined, mind-event? And what has that to do with freedom?

The difficulty can be put forcefully in this way. Suppose that the physiology underlying sneezes is indeterministic in the way that we have postulated the physiology underlying decisions to be indeterministic. One gets a certain feeling that a sneeze may be on its way, and one does not in fact know whether it will come to anything or be dissipated; and it is not physically determined whether it come to anything or be dissipated. In a similar way one knows one may decide to divorce one's wife, or one may decide not to, in the near future; and *ex hypothesi* the decision is not physically determined. The sneeze, we should say, is random; the decision, we should say, is free. But where is the difference between them? In virtue of what would we hold a man responsible for his undetermined decisions, but not for his undetermined sneezes?

Or to put it another way, we earlier maintained that 'I could have decided otherwise' entails that it was naturally possible that I decide otherwise. What is the semantic residue after the *implicatum* here has been subtracted from the *implicans*?

It is to this problem that chapter VI will be devoted.

V

HEGEMONY

(i) Proposal of the theory

In this chapter we introduce a new notion, *hegemony,* which is a feature that can operate in any series of events which is capable of bearing two different levels, or systems, of description. This notion will be the key to solving the second problem put forth in the last chapter, namely, how can the brain state be dependent on the mind state? I shall introduce the notion by a discussion of other cases of series of events which bear two systems of description than the mind-brain series, hoping thereby to ensure that I am not engaged in special pleading.

The three cases I wish to consider are these: the case of the raising of an arm, which can be described either as 'the arm went up',[1] or as 'such and such muscles and tendons contracted, and such and such relaxed';[2] the case of the motion of a child's toy car with a friction engine, which can be described either as 'the car moved forward' or as 'the cogs and wheels went round in such and such a way'; the case of the motion of a billiard ball which can be described either as 'the ball rolled forward' or as 'the molecules of the ball moved in such and such a way'. I shall refer to the former description in each case as the *macroscopic* description, and to the latter as the *microscopic* description.

Now let us look at the relationship which holds between these two descriptions in various cases. Suppose I raise my arm in an ordinary gesture of waving.[3] Here I think we should say that the arm went up because the muscles contracted etc., but not vice versa. And now take the other case: someone grabs my arm and lifts it up; here I think

85

we should say that the muscles contracted because the arm went up, but not vice versa. I can rely only on my readers' intuitions corresponding to my own to secure this point; they can however perhaps be reinforced by an examination of the other examples.

A child is holding his toy car and running it along the carpet to set the friction engine going. Here I believe we should say of the relationship between macroscopic and microscopic descriptions that the cogs are turning because the car is moving, but not vice versa. But now the child lets go the car and it runs along the carpet; here we should say that the car is moving because the cogs are turning, but not vice versa.

A billiard ball is struck with a cue; here we should say that the molecules are moving in such and such a way because the ball is rolling, but not vice versa. Now imagine the case in which a tiny molecular explosion inside the billiard ball causes the molecules to move in such and such a way; here we should say that the ball is rolling because the molecules are moving in such and such a way, but not vice versa.

What is the nature of the 'because' in each of these cases? It cannot be causal, for then we should have an event causing itself, a notion which is at least puzzling and perhaps impossible. The 'because' seems rather to be explanatory: the context seems to determine that the event under one description is explained by its bearing also the other description; and different contexts seem capable of switching around the *explanandum* and the *explanans* as between the two descriptions. We shall examine shortly just what it is in the context that determines which is which.

Often, then, in a system of events bearing two levels of description the one description has what we might call explanatory priority over the other; that the event occurred under the explanatorily prior description explains its occurrence under the other description, but not vice versa. 'Explanatory priority' is a cumbersome expression; I shall use in its stead the term 'hegemony'.

Hegemony, then, is a property which can be possessed by one of two (or more) descriptions of an event, such that the event's occurrence under the latter description(s) is explained by its occurrence under the former description. We could stipulate also that in a case where each description seems to explain the other(s), no hegemony is to be ascribed. That is, a description is hegemonic if (a) it explains and (b) it is not explained by, other descriptions of the same event.[4]

It is important to stress the point that this notion of hegemony is firmly grounded in our intuitions about what it is proper to say in the examples we considered; it is not a philosopher's fiction. We must not defend the libertarian with a *deus ex machina,* and the notion of hegemony is no such thing: it is very firmly on the scene already. I have but noticed it and named it.

Hegemony, we saw in the examples, is conferred by the context. But what are the mechanics of this? What in the context determines which of the descriptions is hegemonic? It is not difficult to abstract an answer from the examples. In the examples the macroscopic description was hegemonic over the microscopic when the event was caused by another event at the macroscopic level; the microscopic description was hegemonic when the event was caused by another event at the microscopic level. Figure 5.1 depicts the six cases as follows: where → designates causal explanation, and ↑ or ↓ designates hegemony, the hegemonic item being written at the blunt end of the arrow and the dependent at the sharp end.

And clearly the ascription of hegemony to, say, the macroscopic description requires not only that the event be causally explained at the macroscopic level, but also that it *not* be causally explained at the microscopic level. Otherwise each description will explain the other, and, as we earlier stipulated, there will in that case be no hegemony.

We are now able to state what is at any rate a sufficient condition for hegemony: a description of an event is hegemonic over others if there is a causal explanation of the event which directly renders the event under that description, and there is no causal explanation rendering it directly under one of its other descriptions. 'Directly' here means 'at the same level' (e.g. macroscopic). Thus the causal explanation 'the child pushes the car' directly renders an event under the description 'the car moves', and indirectly under the description 'the cogs turn'.

Of course the general application of this notion of hegemony would require a way of determining which descriptions of events belong to which level; I have chosen examples in which our intuitions upon this matter are tolerably definite. We need not, however, for our purposes enter into the general problem of assigning descriptions to levels, for our intuitions for distinguishing mental (macroscopic) from neural (microscopic) descriptions are very strong indeed.

We can broaden the above statement of sufficient conditions for hegemony by observing that the hegemony-conferring explanation need not be causal. If the occurrence of an event 'makes sense' under only one of its descriptions, then that description is hegemonic over the others. The point can be established with an example.

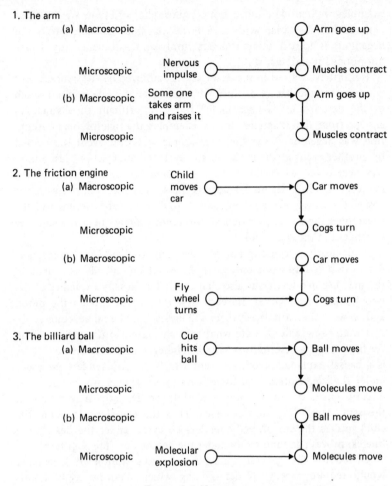

Figure 5.1 Hegemony

If we can speak of banknotes as having a front and a back as we speak of heads and tails of coins, then the following are the fronts and the backs of some denominations of Canadian banknotes:

Denomination	Front	Back
$2	Her Majesty	Eskimos hunting
$5	Sir Wilfrid Laurier	Fishing boats
$10	Sir John A. MacDonald	Oil refinery

Suppose that there is one of each of these notes on the table, and I say to someone that if he brings me one of them he can keep it. He brings me the oil refinery note. Why does he do this? Not 'in order to keep the oil refinery note';[5] not 'in order to keep the Sir John A. MacDonald note'; not perhaps even 'in order to keep the $10 note'; but rather 'in order to keep the most valuable note available' - or something like that. So the event of his bringing me the $10 note is *teleologically* directly explained under the description 'bringing the most valuable note available', and that description is hegemonic. This result accords with our intuitions: he brought the oil refinery note because he brought the most valuable note available.

Or if I say, 'please bring me the note depicting Her Majesty', his bringing me the note depicting Eskimos is only indirectly explained by my request; what is directly explained is his bringing me the note depicting Her Majesty. Suppose I say 'why did you bring me the note depicting Eskimos?' He will reply 'bringing the note depicting Eskimos is bringing the note depicting Her Majesty, and you asked for the note depicting Her Majesty': his account first switches the description in accordance with the identity, and then connects the event, under that new description, with some other event, my request.

So that we can say that a description of an event is hegemonic over others if there is an explanation of the event which directly renders the event under that description, and there is no explanation rendering it under one of its other descriptions.

There is one further complication to be added to this account of hegemony. In a case where there are explanations of the event under several of its descriptions, but only one of them renders the event as unique, the description under which the event is rendered as unique is hegemonic. For an example of this we can add another banknote to the array, a $20 note, which depicts Her Majesty on the front and a scene in the Rockies on the back. Now if I say, 'Bring me a note depicting Her Majesty', and my lieutenant says 'Bring him the note depicting the Rockies', and the man brings the $20 note, the hegemonic description of that event is 'bringing' the note depicting the Rockies', for the explanation 'you asked for a note depicting Her Majesty' would have explained two different events, bringing a $2 note or bringing a $20 note, but the other explanation would explain only the event which actually occurred.

We can take this complication one step further by saying that if an explanation renders an event under a description as unique, and another explanation renders it under another description not only as unique but also as *necessary*, the latter description will be hegemonic. In general: that description is hegemonic under which an event is most strongly rendered by explanation.

In my treatment of determinism in chapter I, I was concerned to argue that neurophysiological explanations of actions would override other sorts, and in the course of that I introduced, in an informal and intuitive way, the notion of *strength* of explanation. And now that notion occurs here again, as essentially bound up with the idea of hegemony. It is time, perhaps, to gather together these pieces of theory and to state the whole in a straightforward way.

An event which occurs usually bears a number of different descriptions. Some of these descriptions will be logically related and others not. An explanation of an event, any sort of explanation, renders that event under one of its descriptions. Explanations are linguistic in the sense that they latch onto a description of an event, not the pure event itself. It will often arise that an event can be explained in various ways, each of these explanations latching onto the event under a different one of its descriptions.

Explanations differ in strength. The weakest merely render an event. Most explanations are of this type: they give reasons why an event happened, but those reasons would also have been reasons for some other event, incompatible with the first, to happen. That is, explanations which merely render an event give conditions but not sufficient conditions for its occurrence. They give, if you like, sufficient conditions for a disjunction of possible explananda. Such weak explanations are often improvable.[6]

Explanations of middle strength render an event uniquely: they give sufficient conditions for an event under that description to occur, rather than conditions which are sufficient merely for a disjunction of kinds of event. At their best, teleological explanations are of this sort.

The strongest degree of explanation is that which renders the explanandum as necessary. The most typical case here renders the explanandum as naturally or physically necessary. If there are degrees of strength of necessity, then there are degrees of strength within this third, strongest, category of explanation. It follows from all this that a teleological explanation can be the strongest or overriding explanation of an event only in cases where an explanation which renders it as necessary is not available.

Finally, the description under which an event is most strongly rendered by explanation is hegemonic over other descriptions under which it is rendered less strongly, or under which it is unexplained.[7]

(ii) Systems of alternating hegemony

We have now the apparatus for imagining, abstractly, a sophisticated system of series of events bearing two levels of description, the descriptions on one level being hegemonic sometimes and those on the other level being hegemonic at other times. Figure 5.2 is a diagram of such a system, where the vertical arrow points from the hegemonic description, and the horizontal arrow indicates a unique-rendering explanation; the vertical line indicates correlation without hegemony.

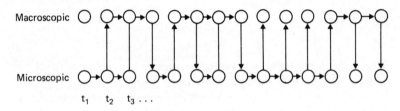

Figure 5.2 A model for alternating hegemony

This model would be particularly apt for depicting the relationship we like to think holds between mind and brain; without any out-and-out causal interaction between the two, it none the less allows the state of the one to be the reason for the state of the other, and vice versa, at different times.

It would, for example, perspicuously represent the difference between spontaneous or stream-of-consciousness thinking and deliberate attentive thinking. It is plausible to suggest that when, for example, we walk in the country and allow our thoughts free rein, it is the neurology beneath them that, according to its own laws, directs their course. On the other hand when we force our thought onto some track, to perform a deduction, or to solve a problem, the stream of mental events is directed by some mental (logical) laws, and the mental descriptive level is hegemonic: here the mental descriptions drag the neural descriptions about according to the laws of sequence which belong to the mental; in the case of free associative thought the neural descriptions drag about the mental descriptions according to laws of sequence (causal laws) which belong to neuralia – if I may put the matter in this

91

picturesque way. These two cases would be depicted as in Figure 5.3.

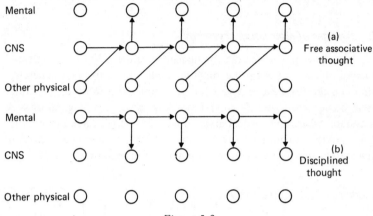

Figure 5.3

This diagram may seem misleading because of the convention that no sets of horizontal or oblique arrows are drawn except to represent sufficient conditions. The sort of indeterminism we have imagined is indeed one which denies that there are sufficient conditions of the neural sort for some neural events, but it does allow that there are sufficient conditions for a disjunction – a shortish one – of next neural events. We might represent this state of affairs as in Figure 5.4.

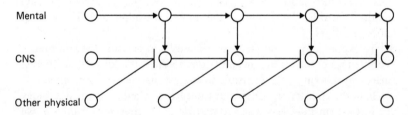

Figure 5.4 Indeterminism in the CNS

When the physical conditions for a mental/neural event are not sufficient, the mental conditions may be sufficient, and in that case the mental description has the hegemony.

It is to be underlined that for the model of alternating hegemony to be applicable to the case of brain and mind there *must* be neural indeterminacies, for unless there are the mental description will never have hegemony. (Here is another reason why the denial of determinism in neurophysiology is necessary to the defence of libertarianism that I am advancing.)

But of course none of this *need* be the case to account for the phenomena of mental life. The Epiphenomenalist model, according to which the neural description always has the hegemony, can perfectly well be the correct one. We have not shown that the model of alternating hegemony does or must apply to the mind and brain; all we have shown is that and how it could. But clearly, for the libertarian, the potential benefits of this model are very great.

(iii) Decisions

But to make use of the notion of hegemony to solve his second problem the libertarian must show that in the case of a decision it is the mental description that is hegemonic. For the second problem was that of resisting the idea that what happens in a decision is that a random (partially determined) neural event brings it about that the *decisum* is what it is. If then the libertarian can find a way of arguing that the mental description of a decision is hegemonic over the neural, he can claim that, on the contrary, it would be more nearly true to say that the decision brings it about that the neural correlate is what it is.

Can he claim, however, that in the case of a decision the mental description is hegemonic? The prospects look very bad, for surely a decision is *par excellence* the mental event for which there are no sufficient conditions of a mental sort, for which, typically, there is no explanation which renders it as unique, let alone as necessary. Suppose the agent decides to divorce his wife; the mental description of the event is ' "divorce wife" appears as *decisum*'; but as far as any mental causation goes there were not sufficient conditions for the event under this description: the contradictory *decisum* could as well have appeared.[8] It looks as though a decision will be an event which is undetermined at both the neural and the mental levels; if that is the case neither description will have the hegemony. For the libertarian this result would not be very satisfactory.

At this point it becomes tempting to make a move which is reminiscent of some forms of compatibilism, and say that a free decision is one which *is* rendered necessary by an explanation. One famous such line is that a free decision is one which is rendered by a string of reasonings;[9] another one, suggested by Wiggins,[10] is that a free decision is one which conforms to the general story of the development of a man's character.

It is unlikely that the latter line could ever be tightened up to serve as a necessitating explanation of decisions: people's characters notoriously enfold contradictory tendencies; and it is doubtful that the former line could be so tightened: one can find reasons for anything. But even if these explanations could be tightened up to serve as necessitating explanations of decisions, they would not really satisfy the libertarian; for traditionally, and, surely, rightly, he wants to claim that a man can freely decide irrationally, or that a man can freely decide to turn over a new leaf and remodel his character.

But there is a move which the libertarian can make to give hegemony to the mental description of the decision. And it amounts to offering a kind of necessitating explanation of the mental event, the decision, but with none of the difficulties which beset the two explanations we have just considered. The move is a valid one on the surface of our conceptual system, I believe, but if it is to be used here it veritably shrieks for explanation at a deeper level. I shall offer the move as a surface move now; the next chapter will amount to an explanation of it.

The mental description of the event which is an agent's decision to divorce his wife is, say, ' "divorce wife" appears as *decisum*'. What explanation renders this as a necessary occurrence? The libertarian can reply: what explains the occurrence of the event under that mental description is precisely that the agent decided to divorce his wife. Given that the agent decided to divorce his wife, it was inevitable that 'divorce wife' appear to him as *decisum*.

This move may seem short-sighted: it may seem merely to shift the problem one stage back, inviting the question, 'what explanation necessitates the event described as "he decided to divorce his wife"?' We shall see in the next chapter how this problem is to be circumvented.

VI

THE RANDOM AND THE FREE

In the previous two chapters we have shown how the state of the brain may sometimes lack sufficient physical causal conditions, and we have shown that at those times the brain state may be dependent on the mind state and not vice versa. Further, we have been able to show this while adhering to a fairly conservative theory of mind, the Identity theory. At the very end of chapter V we outlined how, in the case of decisions in particular, the brain state might be dependent on the mind state – though here the argument was somewhat promissory.

We have now to confront the third of the problems obstructing the libertarian, the problem of finding and describing a *tertium quid* between randomness and causal determination, for freedom seems identical with neither. The libertarian of course cannot stomach the notion that freedom is compatible with causal determination, so he will want to see it as a species of randomness, or as we might rather say, of uncausedness. His problem then is to identify its differentia. Freedom cannot be just plain uncausedness: we have no inclination to describe the uncaused meanderings of electrons as 'free'; what qualification must be added to uncausedness to produce freedom? At the end of chapter IV we put the question in this way: 'he could have decided otherwise' entails that it was naturally possible that he decide otherwise; what is the semantic residue after the *implicatum* here has been subtracted from the *implicans*?

It is worth noting that this problem in the concept of freedom transcends all the mind-body worries that have preoccupied us so far. It would arise even in a world of disembodied minds from which all causal necessity was wished away. Such a disembodied mind would be the

95

subject of many events, all of them uncaused. Some of these would perhaps be free decisions. Even for the disembodied mind the question arises, what is freedom more than mere uncausedness? - for surely it must be more.

I think that this problem is very deep and very difficult; it is difficult even to state the problem in a clear way. It is not to be solved by superficial tricks or technical bravura, but by patient and careful thought. Indeed, I should remark that the discussion which follows is at times distressingly abstract and obscure; I can only say that if I saw any alternative I would take it.

The libertarian's problem here is an instance of a general issue which has been a source of recurring trouble in the history of philosophy - call it the issue of the causal peculiarity of human actions. There have been three broad kinds of view about it, the notion of self-determinism or agent causality, the notion of activity, and the notion of purpose. It will be useful to pursue a solution to the libertarian's problem under these headings.

(i) Self-determinism or agent causality

The idea of self-determinism may be said to have originated with Aristotle. Everyone is familiar, of course, with the suggestion in the *Nicomachean Ethics* that a voluntary act is one whose source (archê) is within the agent, whereas an involuntary act has its source outside the agent.[1] On its own, of course, this remark does not really take us very far; it cannot be said to amount to any very complete or even clear theory. A rather more direct picture is given, however, in the *Physics,* Books VII and VIII.

Early in Book VII Aristotle says that everything which is in motion is moved either by itself or by something else.[2] This has the air of a fairly radical dichotomy, though it is still not a perfectly clear dichotomy, for Aristotle here as often is niggardly with examples. What is meant by a thing's being moved by itself? Acquaintance with Aristotle's physics and cosmology suggests that, for example, fire moving upwards is self-moved and earth moving downwards is self-moved. In the *de Caelo* he remarks that simple bodies are those which possess a principle of movement in their own nature, such as fire and earth.[3] But if the movement of earth downwards is a prototype of human self-movement and so of the voluntary act, Aristotle's theory of the voluntary is distinctly unappealing.

In Book VIII of the *Physics,* however, the matter is set out more copiously and more subtly. There he proposes to cross-divide the distinction between the self-moved and the other-moved with that between natural and unnatural motion. It turns out though that all self-motion is as such natural motion.[4] So only motion derived from something else may be natural or unnatural. Now the motion of earth upwards is unnatural motion derived from something else, and – this is the instructive point – the motion of earth downwards is not self-motion but natural motion *derived from something else.*[5] 'For self-motion is a characteristic of life and peculiar to living things.' And then we are offered what amounts to a criterion for self-motion: 'if a thing can cause itself to walk it can also cause itself not to walk.' We should only call the upward movement of fire self-movement if fire had also the power to stop itself from moving upward. That is, there is bona fide self-movement only when the thing in question has the power to move or not to move with the movement in question.

Now Aristotle would, I think, have great difficulty in saying what it is that causes the natural movements of the elements.[6] But that is perhaps not important for our concern here, which is to understand the character of self-movement. If we attempt to take the concept of self-movement quite seriously, that is, as a radical concept which is not ultimately reducible to movement derived from something else, we are quite stymied. For think of an object which suddenly moves by itself, that is, which is not caused to move by anything else. We should say that it moves spontaneously; but spontaneity is for Aristotle only an epistemic category, not a real one:[7] things which are said to move spontaneously are in fact caused to move in a conventional way. If real spontaneity is not open to him, how can he think that self-movement is in any radical way different from movement derived from something else?

The disappointing answer is, of course, that he doesn't. In the very chapter of the *Physics* that we are considering he makes it clear that self-movement is possible only in objects which are composite, that is only where the object has parts such that one part is the mover and the other part the moved.[8] Now one might have hoped that once one had discomposed the self-mover into the moved part and the mover part, the latter would exhibit, shall we say, pure self-movement (as well as movement of another object, the moved part). But no. Self-movement is possible only for composites, and simples cannot move themselves.[9] It seems that when it really comes down to it

Aristotle cannot identify the mover with the moved: self-movement is not a strict category but a sort of cheat. No object ever really moves itself: all that can happen is that one part of an object moves the rest.

That being so, one wants naturally to know the character of the movement of the mover part of a so-called self-mover. It cannot be self-movement (unless of course the mover part can in turn be dis-composed into a mover and a moved – but ultimately such discomposi-tion will not be possible and at that time the question of the present paragraph will arise). Nor can it be real spontaneous movement, for that is not accommodated in Aristotle's physics. The remaining and dis-appointing possibility is that the mover part of a so-called self-mover is moved by something outside itself. It results that self-movement is but a special case of movement derived from something else: a self-mover does not derive its movement directly from something else, but one of its parts does, and this part then gives the movement to the so-called self-mover.

This rather abstract argument seems to be confirmed by a striking but little-noticed passage in the *Eudemian Ethics*. In Book II, chapter 8, Aristotle is discussing some of the complexities of incontinence, and he passes the remark that the parts of the soul may be said to act from compulsion and involuntarily, but the soul as a whole acts voluntarily[10] (compare, the parts of the animal derive their movement from some-thing else, but the animal as a whole is a self-mover). This pronounce-ment seems arbitrary, to say the least.

Aristotle's distinction, then, between the self-moved and the other-moved is not a radical one but a superficial one: the self-moved turns out to be a species of the other-moved. We might summarize it in this way: the straightforwardly other-moved has the immediate cause of its movement outside itself; the so-called self-moved has the immediate cause of its movement within itself, but the less immediate cause of its movement outside itself. There cannot be self-movement without other-movement.

It is perhaps not useful to us to pursue this subject with Aristotle, for his conceptual apparatus surrounding efficient causality was much less refined than our own. He has, for example, no notion of incomplete or partial causation – an efficient cause is for him a sufficient cause – and we cannot hope to make sense of the causal ancestry of human actions without such a notion. We need to be able to say that a certain desire was an incomplete cause of a given act in the sense that it was a sufficient cause of a disjunction of possible acts, but not in itself a sufficient cause of the disjunct which actually occurred. However all that may be, enough has been said to show the utter inadequacy, for the libertarian, of Aristotle's account of self-movement. For his notion of self-movement is such that it would be entirely compatible with universal causal determinism. A given human act would count as voluntary and a case of self-movement provided only that the deterministic chain of causes leading up to it have its last links occurring within the human being. A remote-controlled model airplane would count as a self-mover for Aristotle,[11] and its flight would be voluntary; and clearly there can be remote-controlled model airplanes in a causally deterministic universe.

Of course the fact that Aristotle did not allow real spontaneity makes it unsurprising that his general view of agency and the voluntary is a compatibilist one. We ought now to turn to a thinker who, while allowing at least the possibility of real spontaneity (incomplete caused-ness) wants none the less to see something like self-movement as the peculiar character of human or animal action. That is, we ought to look at a theory which makes a more radical distinction between self-movement and other-movement than does Aristotle's.

Richard Taylor, in his book *Action and Purpose,* proposes what he calls agent causality. According to this theory the immediate cause of a human act is the agent, not some event or other. He argues that the idea of an agent as cause is the original and more natural causal conception; the theory that only events can cause events is a new and less natural theory which there is no compelling reason to accept.[12]

This proposal is an elegant and appealing one. It is plausible to suggest that the libertarian's embarrassment at not being able to find a *tertium quid* between the random and the causally determined stems from his unwitting subscription to the view that causality is properly a relation only between events. A decision is an event. We ask what caused it and the only possible answers seem to be either that another event caused it – in which case we are not free, or that nothing caused it – in which case we are not free but in the bondage of randomness. Events indeed are either random or else caused by other events; there is no sense at all to the idea of a self-caused event. On the other hand if the libertarian can loose himself from the grip of the idea that only events can cause events, then he can perhaps admit that sometimes objects cause events. And then when he asks what caused a given decision he would no longer be caught between the equally embarrass-ing alternatives that another event did or that nothing did – it just happened. He can find a middle ground between determinism and randomness by saying that the decision was caused by an agent.

Thus far, of course, this move is a surface move, and it needs defence from the obvious and immediate objection that any instance of agent causality can readily be discomposed into a series of events. We are confronted, that is, with an agent's decision and we ask what caused it. We propose the subtle answer that it was caused by no other event but neither was it caused by nothing; rather it was caused by the agent: no other *event* makes the decision occur; it is the agent that makes the decision occur. But surely, the objection runs, this agent is himself just a bundle of events and states; surely what makes the decision occur is just the collective force of all the events and states which constitute the agent. Agent causality was to save us from the embarrassments of event causality; but it is a kind of fool's gold: agents, on analysis, turn out to be just a cluster of events and states: merely a locus of a disturbance. The distinction between event causality and agent causality is not a radical one; agent causality turns out to be just a special case of event causality after all.

It will not do, I think, to resist this objection by claiming that agent causality is conceptually more basic, because an older notion, than event causality, and thus that if any analysis is to be done it is event causality that should be analysed out into agent causality and not vice versa. Age in concepts is not necessarily a virtue, and event causality is a scheme which is simple and which claims, plausibly, to be universal. For these reasons alone event causality seems more basic than the other. To resist the objection we must rather show that instances of agent causality are not capable of being analysed into event causality. Is this possible? Can we stop the analysis of an agent into pure event causality? Can we say that a human act or decision is an event which is caused by an agent and which is not caused by other events? Under what conditions might there be such bona fide agent causality?

The first condition for agent causality, that the agent not be analysable into a bundle of events and states which are sufficient conditions for the occurrence of the act, is met of course in the libertarian's scheme of things by the insistence on indeterminacy in the nervous system. It is the events in the nervous system which would constitute the most detailed train of causally linked events leading up to action. If this train of sufficient causes is broken just before the action (or decision), then the proposed analysis of the alleged agent causality into event causality does not succeed. The events in the agent leading up to the action do not amount to sufficient conditions for the action.

The second condition is a matter of much more difficulty. Given that a certain act or decision was not caused by preceding events, why should we say that it was caused by the agent instead of saying that it simply was not caused at all, it was random? What would be the difference between an event's being caused by an agent, and its not being caused at all? We are back to our old problem: what is the difference between randomness and freedom?

One thing which is clear is that there can be no such thing as agent causality – at least the radical kind of agent causality that the libertarian wants – if one adheres to a mere regularity view of causation. Agent causality is, precisely, irregular. So that if there can be any content at all to the libertarian notion of agent causality it must be something other than regularity; it must be the idea of 'power' or one of its congeners. Now I do not want to open the whole debate about the nature of cause just here; I want only to point out that the libertarian who seeks to solve his final dilemma by invoking agent causality is thereby committed to postulating causal powers or something of the sort. It is traditionally thought that causal powers and their exercise are not observable ingredients in the world, and so their status in a metaphysical theory must be strictly that of postulates. If this were true, then libertarianism would not be a possible position for the empiricist who is ontologically austere.

The libertarian, then, must say that what it means for an agent to cause an event is (a) for that event to fail to be explained by causal links to preceding events and (b) for that event to be linked to the agent by a causal power. It is the presence of (b), the exercise of causal power, that differentiates agent causality from mere uncausality, freedom from mere randomness. But if indeed causal powers and their exercise are not observable ingredients in the world, then we seem to have the highly embarrassing result that we can never know when they are present and when not: we can never know when we have a case of agent causality and when we have a case of mere uncausedness. This will scarcely do. To avoid this awkward charge the libertarian will have to align himself with those who argue that causal powers are indeed somehow detectable and that there is thus a way of telling those occurrences in the world which are merely uncaused from those which, while being uncaused by other events, are none the less caused by agents. He must argue that there is reason for saying that human beings cause their decisions but that electrons do not cause their random meanderings. In the next section we shall discover some reasons for saying this.

Before we get to that, though, there is a small worry to be squared away. At moments when one churns over this very abstract subject it may seem that in being asked to believe in radical agent causality or self-movement one is being asked to believe in something which is strictly inconceivable like – at worst – an event which causes itself. We must examine the precise extent of the logical oddness in which we are involved here.

Events are typically associated with objects, and their occurrence can normally be expressed as an alteration in the features that are attributed to an object. Doubtless many events include changes in several different objects, but that point need not disturb us here. Now when an agent makes a decision, that is an event, and the object to which that event belongs is the agent: the decision event amounts to some changes in the features of the agent. We might say that the decision is an event which occurs to the agent. But in a case of agent causality we have the further point that the agent *causes* the event: the agent causes the event to himself.

Now presumably we shall want to say that the agent's *causing* the event is also an event. We seem then to have two events, the decision which is an alteration in the agent, and the agent's causing that alteration. At once there looms a vicious regress. It can be forestalled only by saying that these apparently two events, the decision and the agent's causing the decision to itself, are in fact one and the same. This is the logical oddness to which we are committed. How intolerable is it?

Indeed it would not be possible to allow the idea of a self-causing event, that is an event which is the same event as the event which caused it. Event causality, if you like, is nonreflexive. But that is not what is required here; what is required here is subtly but importantly different. We do not require that an event be the same event as its cause, but that an event be the same event as its being caused.

This sounds like the most grotesque logic-chopping of the Schoolmen, but I believe that what it proposes is both important and true. We can show that our conceptual system will indeed accommodate the oddity of an event being the same event as its being caused. It is precisely this that is the intriguing feature of so-called basic acts, acts which are done without the agent's needing to do anything else in order to do them. In a case where I raise my arm 'I make my arm go up' and 'my arm goes up' are one and the same event. I cannot distinguish a separate making my arm go up from the simple event of my arm's going up. It was perhaps a hesitancy about allowing that an event could be the same event as its being caused that led earlier theorists to postulate ubiquitous volitions. Volitions are not often there, and they only postpone the problem when they are there. We need rather simply to admit that an event and its being caused can be one and the same event; our experience of action suggests that they can be. I conclude then that agent-causality is logically peculiar but certainly not obviously impossible. Its peculiarity should come, of course, as no surprise to the libertarian: he is already committed to the view, or the hope, that free agency is deeply different from ordinary causality, and seeks only to know in what its difference might consist.

It will be wise in a moment to recapitulate and redescribe the state of play in the libertarian's quest to find a stopping-place between the random and the free. But first I should like to offer a kind of passionate apology for the obscurity and abstractness of the preceding and some following pages. At times these matters seem so difficult – not only to solve but even to talk about in a clear way – that one is tempted to join up with the compatibilists and have done with it. I wish the subject were more straightforward. One must however take up philosophical problems where one finds them. The libertarian's problem about mere randomness is scarcely explored in the literature: it is a problem that has hardly yet been clearly stated. All that we have is a definite state of discomfort, a knowledge that there is something fishy lurking in the issue. The only philosophical procedure that is possible in such a situation is to probe and toy in the hope that things will come clear in the end.

The state of the libertarian's argument so far, then, is this. If one insists that event causality is really the only kind of causality, and that so-called agent causality or self-movement is just a special case of this, then there seems little hope for the libertarian to find a *tertium quid* between causal determination and uncausedness. If however one is willing to admit that some events which are not caused by other events may none the less be caused by an object or agent (which object or agent cannot, of course, itself be discomposed into events), then a middle ground may be found. The middle ground between being caused by events and not being caused is being caused by an object. This then is a sketch of how we can get freedom out of mere randomness. If we can bring ourselves to allow that agent causality is a perfectly bona fide type of causality, radically distinct from event causality, then we shall have the scheme shown in Figure 6.1.

	caused by event	not caused by event
caused by object	overdetermination	freedom
not caused by object	normal causal determination	mere randomness

Figure 6.1 Conceptual map:
agent causality and event causality

The genuinely random is that which is caused neither by preceding events nor by an agent; the free is that which is caused by an agent but not caused by any events (including events in the agent); most events in the world are caused by preceding events and not by an agent.

Whether there could be radical agent causality in a case where there was also event causality is complicated. Certainly there could not be if the event causality in question ran through the agent in such a way as to make his causing the event itself caused by other events (as in Aristotelian self-motion). But if the event causality worked so to speak from the outside, then perhaps both sorts of causality would occur and we should have a case of overdetermination.[13]

To get this solution to his problem, then, the libertarian must satisfy himself as to the possibility and the propriety of introducing a kind of causality radically different from event causality, not indeed displacing event causality but supplementing it. He must deny not that event causality often occurs, but that the event causality scheme is universal. It is not really in its overt claims for causal determinism that the scientific world-view conflicts with human freedom (for those claims are no longer always upheld) but rather in its much more covert claim – its undeclared and unexamined presupposition – that the only causality is event causality. What the libertarian must do, it seems, is uncover and challenge this presupposition of science. His debate then is whether at the bottom rung of analysis the only possible causal relationship is between events, or whether there can also be causal relations between objects and events at that bottom rung. His project is to introduce agent causality as an elementary sort of causality.

An initial difficulty facing him here was the suspicion that agent causality or radical self-movement was somehow an incoherent idea, requiring reflexivity where there just couldn't be any. We have shown that the libertarian is not here committed to a self-caused event, which would be impossible I think, but to an event's being the same event as its being caused, which seems odd, certainly, but not impossible. Indeed it seems that that oddness is precisely the oddness we find in so-called basic acts. The oddness which the libertarian must admit is an oddness which we all admit anyway.

That initial worry thus squared away, the libertarian might want to go on to plead that our ordinary everyday conceptual scheme already accommodates agent causality. We naturally speak of agents as causing events: the man made his arm go up. It is true that we speak in this way and that our ordinary conceptual scheme accommodates agent causality; but it doesn't follow that that scheme will require or even allow agent causality to be a *basic* kind of causality, a kind of causality which cannot be analysed into event causality. To put it another way, the libertarian's undertaking is not just to defend the idea of agent causality; it is to defend the idea of what I have called *radical* agent causality. The surface idea of agent causality is indeed familiar to us; it is easy enough for a compatibilist to make that surface notion good as a special case of event causality: that was just what Aristotle did. The libertarian's task is much harder; he must make the notion of agent causality good as a basic or irreducible kind of causality. So an appeal to the fact that the notion of agent causality figures already in the quiver of concepts with which we organize our world does not advance his real enterprise, which is to show that the notion might be a *basic* notion of causality.

What, then, are the obstacles to seeing agent causality as a basic kind of causality? Why should we not just allow that while events are often caused by preceding events, sometimes they are not caused by preceding events but by objects – by agents? For a long time this idea seemed impossible to me because I could not conceive how an object considered apart from its states or changes within it could be said to *cause* anything. But this of course was to seek to understand radical agent causality on the model of event causality – that which *ex hypothesi* it is not possible to do. It was to refuse to give up event causality as the basic model; it was to suppose that something which is merely a deeply-lying ingredient in our scientific view of the world was a necessity of thought. I mention this to emphasize how firmly we are in the grip of the idea that event causality is basic and universal: it is as I said before a largely unexamined presupposition of our science.[14] Such unexamined presuppositions can be extraordinarily difficult to shed, so difficult that shedding them may sometimes seem to be logically impossible.[15]

The libertarian, then, is proposing a kind of causality which is not capable of being analysed into our normal kind of causality, event causality. It may seem that he is thereby wantonly introducing mystery into the world. I think that this charge is correct, but perhaps it can be rendered less damaging by the following anodyne consideration. He is not so much introducing mystery into the world as introducing more mystery into the world: the event causality with which we seem so comfortable is itself unfathomably mysterious, as any glance at a freshman metaphysics text will show. He is not introducing mystery alongside clarity, but mystery alongside mystery. Still, one mystery is better than two, and it is one of the debit points for the libertarian scheme that it must propose additional primitives. The libertarian however thinks that the benefits here outweigh the costs.

However primitive agent causality may be, though, the libertarian must surely tell us in what the causal link between agent and event consists; that is, he must say what the difference is between a case where an object is subject to a random event and a case where the object causes the event to itself. It is not perhaps necessary to the coherence of his theory that that in which the causal link consists be detectable, but clearly his theory would be more attractive if the causal links were somehow detectable.

Of course there has been since Hume a problem over detectability with ordinary event causality. There the standard view has been that we cannot detect or even conceive the necessity which links a cause to its effect; at most we can get a clue to its presence from the universality or regularity with which like effects ensue upon like causes. This line will not be open of course to the defender of radical agent causality, for here there is, as we have said, no regularity. It is not true that whenever you have a certain object you get a certain event; the very idea is nonsensical. It is of course true that whenever an agent causes an event that event occurs, but that is a trivial point because the agent's causing the event and the event's occurring are one and the same event. Agent causality is in no way lawlike.

We can bring the libertarian's problem here into sharper focus by thinking of two cases, a man who makes a decision and an electron that suddenly moves to a new energy level without being caused to do so. For the libertarian neither of these events is necessitated by preceding *events*; but he wants to say that the man causes the decision to occur (the *decisum* to appear, perhaps) in himself, whereas the electron does not cause the movement to itself: the movement merely happens.

What, though, would this difference consist in? What, so to speak, is going on in the case of the man that is not going on in the case of the electron? To answer merely 'agent causality' is to be unpalatably cryptic; on the other hand we must remember that the libertarian is proposing agent causality as a basic kind of causality, as a primitive. Certainly if in our question about what is going on in the case of the man we are looking for some events in him, then we are guilty of the narrow-mindedness which I described a few paragraphs back; we are seeking to understand a proposal in a way which does not take that proposal seriously; we are seeking to understand a proposal by means of an explanatory scheme which that proposal would reject. We must be clear then, that, in asking the libertarian to tell us more about what is going on in the case of the man deciding, we are not insidiously gripped by the assumption that events in the man are the only thing that would constitute a satisfactory answer to that question. Is there anything that the libertarian can say which, while telling us something more than just that agent causality is going on, would however not destroy his position by offering some events in the agent as what constitutes the agent causality?

He can I think say that what is going on in the case of the man which is absent from the case of the electron is the exercise of an *active power*. The difference between the man and the electron is that the man has a power to decide whereas the electron, though it moves, has no power to move. It has perhaps the passive power to move, but not the active power: it has the power to undergo movement but not the power to move itself.[16] We have agent causality, then, when we have the exercise of an (active) power that the agent possesses; we have mere uncausality when the object which moves lacks the (active) power to do so, or has that power but is not exercising it. The libertarian, to fill out his position a little bit, can thus offer the notion of power as a notion in whose terms to understand the otherwise mysterious distinction between radical agent causality and mere uncausality.

The next question which arises for the libertarian is whether these powers are transcendental postulates in his theory, or whether they are detectable. On this subject of detectability of power there has been a good deal of debate. My own view is this. I think that by direct experience we can detect the exercise of our own active powers and so we can infer their existence; we can only infer both their existence and their exercise in other persons or objects.[17] We infer that other people and animals have such powers and we infer that electrons do not. This inference works largely by analogy; it may be mistaken.[18]

I do not want to embark on a general defence of this view that we can detect the exercise of active powers in ourselves, but only to answer one interesting, apparently forceful, but finally insubstantial argument it gives rise to. The argument works in this way. We ask how it is that we detect the exercise of an alleged active power as distinct from the mere playing out of something we undergo. The obvious answer – which is not circular – is that in the exercise of what we call an active power we feel active; when we merely undergo something, we feel passive. That is the important experiential difference between, say, a decision and a sneeze. (There are, incidentally, very few other experiential differences between those two.) So our detecting the exercise of an active power in ourselves amounts to this: we feel active, not passive.

Now the libertarian is ultimately under the requirement that he say what an active power consists in – not just what it feels like; his problem is a metaphysical one. And when we try to say what an active power consists in, our mind goes blank in a familiar way. It seems as though this detective clue, feeling active, is a clue to nothing imaginable. We just cannot imagine what an active power would be: any way we turn it it ends up as something we undergo, to which we are passive. The determinist then accuses the libertarian of postulating something unimaginable when he postulates active powers, and goes on to fill out his case by saying that feeling active is really no evidence at all that we are being active – that is, being radical agent causes. Feeling active is a feeling on all fours with other feelings: we can perfectly well undergo it just as we undergo hunger or heartburn. So far from detecting an exercise of real radical agency, we may be detecting just another thing we undergo. Feeling active is scarcely a good clue that we are being active, in the radical sense. The libertarian postulates powers to answer the phenomenon of feeling active; the determinist complains that what the libertarian has postulated is unimaginable, and points out that the phenomenon it was designed to save can easily be saved in a deterministic scheme.

By way of reply, the libertarian must in the first place admit that it is perfectly true that feeling active is not conclusive evidence for being active in the radical sense: that experience is as capable of being illusory as is any other. The libertarian's acquiescence here is part of his general admission that the determinist's position, in an appropriately subtle version, is entirely capable of accounting for all the phenomena of experience. The libertarian is not trying to save that sort of phenomenon, but rather some phenomena of a moral kind, phenomena over which he thinks that the determinist does rather badly.

In the second place, the libertarian must admit to sharing the difficulties about imagining radical agent causality, or real active powers. He does not believe that it is an inconceivable thing to postulate, however, only that it is very difficult to perform the suspension of presuppositions that is required before one can do so.

But there is a sense in which active powers seem to be especially and necessarily unimaginable. The ingredients of the imagination derive ultimately from experience, and in a way we can never *experience* activeness. We can never directly experience ourselves being radically active, because experience is something which we undergo, to which we are passive. Thus we can never know what it is like to be active; all we can know is what it is like to feel as though we are being active. The libertarian can offer this palliative consideration when he is challenged with the unimaginability of real active powers: we cannot imagine them because imagination is based on experience and we can't directly experience them. Our experience of radical activity is always veiled in passivity, just because it is experience. Of course, something can be conceivable, that is, logically possible, without being imaginable, so the libertarian has not undercut his own position here. But he does now seem to be involved in postulating something quite transcendent when he postulates active powers. The determinist may be unimpressed by this and think that if libertarianism involves such abstruse twistings and turnings we had better stay clear of it. At this point libertarianism seems extremely unattractive.

In fact, though, active powers are not a very special case. We ask what the exercise of an active power really consists in, and find that all we can say is what it feels like: we can never get at it, itself. But we are in just the same condition with regard to the external world: we ask what it is really like and find that all we can say is what it seems like. The external world which we postulate can never be experienced in its externality, because experience is internal; active powers can never be experienced in their activeness, because experience is something to which we are passive.

The analogy indeed is instructive, and it offers the libertarian a stick with which to beat the determinist. For the determinist is really no better than a sceptical solipsist. Both of them succeed in saving the experienced phenomena of life, but both of them do so at the expense of holding that that experience is illusory: the one denies that what seems external really is, and the other denies that what feels active really is.

We have strayed somewhat into the subject of the next section; we ought to turn to it properly, now, and make a fresh beginning.

(ii) Activity and passivity

The libertarian was facing the question about the difference between freedom and mere randomness, the question about the difference between the case of the man freely deciding and the case of an electron uncausedly moving. Both of these events are, in the libertarian's view, uncaused (by the usual kind of causality); why should the one be called free and the other random?

One very satisfying way of summing up the difference between the two cases would be to say that the man is active in his uncaused decision whereas the electron is passive in its motion. The decision is something which the man uncausedly *does,* whereas the electron uncausedly *undergoes* its motion. Indeed that seems a precise way of capturing the difference that the libertarian would want to maintain between the two cases.

The difficulty is that thus far the distinction between activity and passivity is but a surface distinction. One wants to know what it consists in. What, really, is activity?

And here again as with self-movement and agent causality, compatibilist interpretations have often been offered and are easy to offer. In a deterministic universe an instance of activity would be an instance of causation by events within the agent, that is, an instance in which the causal train of events leading up to the 'act' passed through the agent.[19] If we consider, not implausibly, that any event belonging to an object which is caused by a preceding event is an event to which that object is passive, then it follows that on the compatibilist model of activity, activity is but a special case of passivity.

We can see such a compatibilist notion of activity very clearly in Leibniz; here is his first definition of action: 'an action is a change in something of which the proximate cause is in itself.'[20] In this definition the presence of the word 'proximate' shows that the kind of activity in question is compatibilist. A non-compatibilist idea of activity would have used 'original' in place of 'proximate'. Here is another passage in Leibniz, from his Fifth Paper, in which we see him very explicitly offering a compatibilist's shallow distinction between activity and passivity:[21]

14. . . . I now come to the objection made against my comparison between the weights of a balance and the motives of the will. The author objects that the balance is purely passive and weighed down by the weights, whereas agents which are intelligent and endowed with will are active. To this I reply, that the principle of the necessity of a sufficient reason is common both to active and to passive things. They need a sufficient reason for their activity as well as for their passivity. Not only does the balance not act when it is weighed down equally on both sides, but equal weights do not act either, when they are in equilibrium in such a way that one cannot go down without the other going up to the same extent.

15. It must also be considered that, strictly speaking, motives do not act on the mind like weights on a balance; it is rather the mind which acts by virtue of the motives, which are its dispositions to act. Thus to maintain, as it is here maintained, that the mind sometimes prefers weak motives above stronger ones, and even sometimes what is indifferent above motives, is to separate the mind from its motives, as if they were outside of it, as the weight is distinct from the balance; and as if there were in the mind other dispositions to action besides motives, by virtue of which the mind rejected or accepted the motives. Whereas the truth is that motives comprise all the dispositions which the mind can have to act voluntarily, for they comprise not only reasons but also inclinations, which come from the passions or from other preceding impressions. Thus if the mind preferred a weak above a strong inclination it would be acting against itself and otherwise than it is disposed to act. This shows that notions which differ on this point from mine are superficial and turn out to have nothing in them, when they are properly considered.

It is clear that such a theory as this is altogether consistent with – indeed it almost commits one to – universal causal determinism of the acts of the mind.

The libertarian of course must find an idea of activity which will not be compatible with universal causal determinism. And at this point he encounters an interesting phenomenon, an interesting parallel with the inquiry into agent causality. In the inquiry the libertarian happily denied that every event is caused by a preceding event, and then found himself battling with a very tenacious residue of determinism, the theory that all causality is event causality. The real problem for him was not the determinist theory itself but rather one of its presuppositions, the event causality scheme. It was easy enough to reject determinism, but to divest oneself of this presupposition was much more wrenching.

In pursuit of a non-compatibilist notion of activity, the libertarian meets a parallel difficulty. In a deterministic scheme activity is but a special case of passivity: passivity is so to speak the basic mode. Now when the libertarian, pursuing activity, denies universal determinism, which is easy enough to do, he is again left with a clinging residue: this time the residue is the idea that passivity is the basic mode. One thinks of a causally undetermined event, an electron's random motion or a man's decision, and one feels it inevitable that finally these both be regarded as something which is undergone by the object in question, the electron or the man. The libertarian dismisses determinism easily enough: its presupposition of universal passivity is much harder to dismiss. Even when one allows that some events are uncaused by preceding events, it seems impossible to give up the idea that the objects to which those events belong undergo them. If such events are not imposed on them by preceding events, then they are imposed on them by randomness.

So that on this front as on the last the libertarian must propose not just the revision of the deterministic hypothesis but also, and with more difficulty, the revision of one of its presuppositions. He must propose that activity is not a species of passivity, that the two are basic modes neither reducible to the other. This of course is a proposal that he must explain: we have a tolerably clear idea of what passivity is - we know what it is for an object to undergo a change; we need however something more about what activity is if it is not just to undergo a special kind of change.

An initial problem which can be envisaged here is that language, or our system of concepts, seems to permit the universal recasting of alleged pieces of activity as passivity. If we say 'uncausedly he made his hand go up' and, following a libertarian line, mean that what occurred was an out-and-out irreducible piece of activity, then it is an embarrassment that we can immediately shift our view, so to speak, and see that exercise of active power as an event that occurred to him. Thus 'it happened that uncausedly he made his hand go up' seems a permissible inference from 'uncausedly he made his hand go up.' That is, while affirming that indeed he actively did so and so, we can also say that one of the events he underwent was the uncaused active doing of so and so.

112

It always seems possible to say that though one *does* an act, one *suffered* the doing of the act: it always seems possible to nest activity in passivity. What this shows, I think, is that the all-is-passive theory lies very deep with us.

An analogy here is offered by the argumentative steps leading to the philosophical position of fatalism or logical determinism. We suppose that Cindy will wash her hair tomorrow, 28 April. That hair-washing is a fact which enters into the constitution of the world's career, and its date is 28 April. But next we go on to argue that if Cindy washes her hair on 28 April, then it is true that Cindy washes her hair on 28 April. So we have an additional fact, the fact that it is true that Cindy washes her hair on 28 April. But this new fact has as its date 'always' because of the eternity of truth. Or put another way, that Cindy washes her hair on 28 April is a fact whose temporal locus in the career of the universe is 28 April; that it is a feature of this universe that Cindy washes her hair on 28 April is a fact whose temporal locus is always. With a few more steps involving historical necessity we get to fatalism, the view that it is already on 27 April a fixed feature of the universe, and so inevitable, that Cindy wash her hair on 28 April. It is always possible to nest a momentary fact in an eternal fact. Just as the truth of a fact is itself a fact, so something's being made by an agent to happen is itself a happening. Unless we are to say that the agent made himself make his hand go up, and follow the infinite regress which thereby begins, it is hard to see how we can avoid allowing that all exercisings of active power are themselves events which an agent undergoes.

We are right, I think, to be suspicious of both these nesting operations and to think that those tricks of thought and language which permit the truth of a fact to be regarded as itself a fact, or the making of an event to happen to be regarded as an event which happens, are metaphysically pernicious. We can, I suppose, block fatalism by distinguishing first-order from second-order facts and refusing to recognize the latter as bona fide facts. That is, we reject the supposition that every true proposition is, or stands for, a fact. Resisting the idea that our doings are themselves events we undergo is more delicate, and drives us to make an important distinction.

We remarked above that events are generally said to belong to objects in that they can be analysed as changes which occur in the list of features belonging to objects. For our purposes we can consider only the simple case of an event which involves changes in only one object.

113

Every such event, then, may in this minimal sense be said to belong to, or to occur to, or to happen to, an object. That is, all events indisputably happen or occur, and of those events that belong to an object we say that they happen or occur to the object to which they belong. Now 'happen to' and 'occur to' suggest passivity; let us call this sense of 'happening to' or 'occurring to' nominal passivity.

There is also a 'real' sense of passivity according to which an event not only belongs to an object but is imposed upon the object; the object is made to undergo it. The normal mode of such imposition, if not the only mode, is causation.

Now I think that expressions like 'happen to', 'occur to', 'undergo' can all be used to express either the nominal or the real sense of passivity, and that in this ambiguity lies the solution to our problem about the nesting of activities in passivities. It provides a harmless way of allowing that the all-is-passive theory is true.

The all-is-passive theory is true in so far as it speaks of nominal passivity: all events (or anyway nearly all) happen to objects. But that theory is false if it speaks of real passivity. An object is really passive to an event if and only if that event is caused to happen to that object. An object is active if it causes an event to happen and if it is not caused to cause that event to happen. (Notice that an active object is also really passive inasmuch as it undergoes the event it causes to itself: the man who raises his arm is really passive to the rising of his arm.) Finally, in events which occur to an object but which are not caused to that object and which the object does not cause to itself, the object is neither active nor really passive: it is nominally passive. In this scheme (Figure 6.2) we have a case of free agency when an object is nominally passive to an event, active in it, and when, of course, the event is not caused by other events. We have ordinary causation when an object is both nominally and really passive to an event, that is, when the event is caused by other events. And we have randomness when an object is nominally passive to an event but neither active in it, nor really passive to it, that is when the event is not caused by other events.

It is important to be clear that in this scheme, while nominal passivity is in a way basic, so also are real passivity and activity: neither mode is explicable in terms of the other, and while both presuppose nominal passivity neither is analytically exhausted by it. Activity and real passivity are equally basic.

The distinction now between the case of the man uncausedly deciding and that of the electron uncausedly moving is that while both

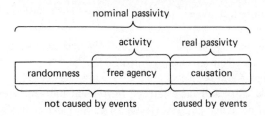

*Figure 6.2 Conceptual map: causation,
activity, passivity*

are cases of nominal passivity, the former is also a case of activity and the latter not. In this way, by the distinction between nominal and real passivity the libertarian renders metaphysically harmless the fact that language and thought seem to permit the nesting of an activity in a passivity. That was, however, but an initial difficulty, and the libertarian has still to say something more about activity, about what it consists in.

In a way we have already done so in the course of developing the theory of nominal and real passivity and activity: an object is active when it causes an event to itself, that is, when it indulges in agent causality. This may seem a disappointment. We entered the discussion of activity to look for an account of what agent causality consists in. The notion of activity itself seemed mysterious and now we venture to elucidate it by reference to agent causality. This is circular, of course, but not viciously so. The libertarian proposes agent causality and activity as basic notions not reducible to event causality or real passivity respectively. It is of the nature of a basic notion not to be analysable; the most we can expect to do is to build up a fairly wide conceptual net around it.

(iii) Purpose

The libertarian has not so far been able to say what radical agent causality or radical activity is, except that they are the same as each other and not analysable as event causality or real passivity. Those paths of enquiry seem shut, and we can turn to another. We laid upon the libertarian the burden of saying what the conditions for agency are – what features an object must have before it can engage in radical agent causality – and that is a burden he has not yet discharged. If he were to discharge it we should have a clearer idea of what agency is for

115

him. We must not expect to be able to state sufficient conditions for agency, but it would help if we had some necessary ones.

We have the case of the man uncausedly making a decision and the case of the electron uncausedly moving; the libertarian wants the man to be an agent and the electron not; he wants the decision to be a piece of agent causality and the electron's movement to be mere uncausality, the man to be active and the electron weakly passive. But why should that be so? For anything that has been said about agent causality or activity so far, it would seem perfectly possible that the electron as well as, or instead of, the man be endowed with that mysterious capacity. True, our normal view is that human beings are agents and electrons not; but the libertarian must not just adopt that view without comment: he must say what human beings have that allows them to be agents and what electrons lack that precludes them from being agents.

One would have thought that consciousness had something to do with it, but it is hard to see why it should, just like that. Consciousness can be impinged upon just as an electron can be impinged upon: why should consciousness also have or confer the capacity for real activity? Consciousness is indeed, I believe, a necessary condition for that capacity, for radical agent causality, but the argument why it should be so is somewhat circuitous.

I think that a necessary condition for agency is purposiveness: a man is said to cause his decision partly because he has a purpose in that decision; an electron merely undergoes its motion and does not cause it, because it has no purpose in that motion. Part of the meaning of 'random' is 'aimless'. All radical agent causality, all real activity, is purposive. That is not of course to deny that human beings sometimes do things on a whim; indeed they often do. But doing something on a whim is always purposive: to gain pleasure or satisfy curiosity. We should never say of an electron that it moved on a whim. Whimsy is purposive, and electrons are incapable of purpose. A necessary condition, then, of radical agency and so ultimately of freedom is purpose: any case of radical agency must be a case of purpose.

But as with agent causality and activity there are many excellent compatibilist accounts of purpose. Cybernetics, the study of goal-directed self-regulatory mechanisms, shows beyond any doubt that goal-directedness or purpose is compatible with causal determinism: indeed it is the science of making such end-pursuing machines. At this point – so near the end – libertarians have sometimes capitulated in the past: it seems very plausible to say that the difference between free

116

agency and randomness is, at least partly, that freedom involves purpose and randomness not; but it turns out that purpose is possible for a deterministic mechanism. If purpose is thus the essential ingredient in freedom, a deterministic mechanism which has purpose can be called free. The libertarian is strongly tempted to succumb.

Our libertarian, however, will not give up at this point. He will admit – it is impossible not to – that a deterministic mechanism can be purposive in a perfectly bona fide way. But he will deny that purpose alone confers radical agency, real activity, freedom. Purpose is a necessary but not a sufficient condition of these. Another necessary condition is uncausedness, that the mechanism in question not be deterministic. We have then two necessary conditions for free agency.

An object is a real self-mover only when (a) the event of its self-movement is not caused by other events, and (b) its self-movement is purposive. We have an instance of radical agent causality only when (a) the event is not caused by preceding events, and (b) the object to which the event belongs has a purpose in that event. There is real activity only when an event belonging to an object is (a) not caused by preceding events, and (b) figures in a purpose of that object.

These conditions are not however sufficient, for it seems that an object could uncausedly undergo an event which served one of its purposes (and even which it knew to serve one of its purposes) without its also being true that the agent caused that event. We have not succeeded here in providing a complete analysis of radical agency, but then we did not expect to: it is introduced by the libertarian as a basic notion. We have however elucidated it.

I wrote above that a necessary condition for free agency was consciousness. We are now in a position to see why this is so. There has been much argument, successful in my view, to the effect that an object can be purposive in its behaviour without being conscious, because purposiveness can be 'reduced to' ordinary causality – in Aristotelian terms, that final causes really work by efficient causality. The libertarian must turn all this argument upside down, however, and point out that while it is possible in principle to reduce purpose to efficient causality, it is possible to do so only on the assumption, usually granted, that the purposive object is deterministic. You can reduce teleology to efficient causality only provided you have an efficient causality to reduce it to. The libertarian however disallows the idea that the operations of a free agent are deterministic – for him there is in a free agent no causal determinism to reduce its purposiveness to.

And since that reduction was the condition for making purposiveness work without consciousness, a free agent cannot be purposive unless it is conscious. Now we have seen that an agent is not free unless it is purposive; it follows that free agents must be conscious.

Let me say all this again. Our ready notion of purpose is such that only a conscious being can have purpose. But it turns out that unconscious beings may be said to have purpose if they are cybernetic engines; part of being a cybernetic engine is being deterministic, being such that every event which occurs to one is entirely determined by preceding events. The libertarian's version of the human body, though, is such that a human agent is not deterministic. It follows that a human being is not a cybernetic engine: he cannot have purpose unless he is conscious. Again, there can be purpose without consciousness only in a deterministic cybernetic mechanism. A free agent is not, for the libertarian, a deterministic mechanism. It follows that a free agent cannot enjoy purpose without consciousness. Consciousness is a necessary condition of purpose in a nondeterministic object. And so only conscious things can be free agents.

(It has sometimes been argued, plausibly, that consciousness is biologically otiose[22] and so that its existence is mysterious. It will be something of a payoff for the libertarian if he can contradict that conclusion by showing, as we have just done, that consciousness is necessary to freedom and purpose, and further that freedom and purpose are necessary to biology. I shall not pursue that interesting subject here.)

Incidentally, it has been usual to restrict free agency to adult human beings; for anything we have said to the contrary here it can belong to children and animals as well. Both children and animals are conscious and purposive, and their nervous systems are not in any significant way unlike those of adult human beings: if neural indeterminism is the case for us it is likely to be the case for them as well. The normal grounds for restricting free agency to adult persons is that only they are rational. I think that rationality has nothing whatever to do with free agency: a crazy man can be just as much a free agent as a sane man. That being so it would seem arbitrary, a sort of metaphysical fiat, to restrict free agency to adult persons. And indeed I have no difficulty in allowing that free agency can belong to children and animals. The embarrassments that seem to flow from that admission will be cleared up when this theory is extended into the matter of responsibility, in the next chapter.

(iv) Theory of mind

An answer to the question about freedom and randomness is now perhaps beginning to emerge. The libertarian must escape the dilemma by postulating agent causality as an irreducibly basic kind of causality alongside event causality. We have explored the possibility of that at some length, and we have woven some of the conceptual network which surrounds the view that there are these basic kinds of causality. There is no question but that we are here involved in a kind of conceptual revision, and proposals for such revision must always undergo cost-benefit analysis. The libertarian will pay anything, of course, to avoid compatibilism; the cost roughly, is that we need two primitives where we had one before. We glimpsed a chance, on the credit side, that the libertarian would be able to solve the baffling problem that consciousness is biologically useless – but that exciting scheme remains programmatic.

We ought to look, finally, at a place where one might have feared large debits of new mystery from the libertarian thesis – the theory of mind. We have been working quite explicitly, until this present chapter, with the Identity theory of mind and brain. One might fear that we have now however got ourselves into a corner in which we need something grander, some sort of undetectable agent-self, like Sartre's transcendental ego,[23] to be the object which causes decisions and so avoids their being random.

The fear would run this way. We negotiated, in chapter V, a way in which the mental description could sometimes have hegemony over the neural description. To secure this hegemony at the moment of a decision, so that the decision's being what it is will not owe just to a piece of randomness in the nervous system, we saw that we should have to find a way in which the mental description of the decision was rendered more strongly by explanation than the neural description. *Ex hypothesi* the neural description, being causally undetermined, can be rendered neither as unique nor as necessary. Can the mental description be rendered any more strongly than the neural?

After the inquiries of this present chapter we can see an obvious way to accomplish this. We have merely to say that the event 'the *decisum* appeared' was caused by the agent in an instance of what I have called radical agent causality. In this way we give hegemony to the mental description.

By now – and this is where the worry begins – it might seem that the

119

'agent' in question, since it operates at the level of the mental descrip-tions, must itself be mental. We seem to be required to postulate an object, the Mind, or Self, in addition to the array of mental descriptions of events belonging to a person with which we have hitherto made do. Moreover, as Hume notoriously pointed out, such a Mind or Self is not discoverable by inner or outer sense; it would have therefore to be transcendent. We seem to have got ourselves into a theory which, while not impossible or incoherent, is much more extravagant than we had hoped. The libertarian wants his theory to be as attractive as possible, however, and it would certainly be made less attractive by the inclusion of an undetectable Self.

However the postulation of a Mind or Self as agent arises through the overlooking of a subtle point. It is true that we need an object of some sort to be the cause of the decision, if we are to have an instance of agent causality. But it is not true that that object has to be a mental kind of object - a Mind or Self; all it needs to be is an object which can bear a mental description. Such an object is a person.

To say this sounds like a tiresomely standard move - almost a ritual move - in the austere philosophical fashion which prevails. But in this case it is a move that is thoroughly justified by the conjunction of several metaphysical doctrines to which we have subscribed.

The first of these is the Identity theory itself, together with the doctrine that events belong to objects. The version of the Identity theory we have defended does not commit us outright to the con-clusion that mind and brain are identical objects; what it says is that mental events and neural events are identical events: certain events bear both a mental and a neural description. Now to what object do these events belong? The only possible answer is a person (or animal, I suppose). It is fundamentally incorrect to separate out mind and brain as distinct objects, for this version of the Identity theory. For how could the event of a neural discharge occur to a mind which was only a mind? And how could a pleasant olfactory sensation be an event under-gone by a brain which was only a brain? The brain and the mind are indissolubly fused into one object, under the Identity theory, because the two kinds of event which occur to them are fused into one event.

(We do, of course, speak of brains as though they were separable objects; that I believe is a mistake which arises from the fact that a person who is looking at a brain is inevitably blind to one whole array of its properties - the mental ones. Of course a dead brain *is* an object in its own right, and indeed it is the very object that we have in mind when we speak of 'a brain'. That is, we speak of both a 'brain' in a jar and a 'brain' in a living person. But they are not at all the same thing. 'Brain' is, if you like, homonymous. Aristotle made precisely this point when he said that a dead hand is not a hand.[24])

Thus although the Identity theory begins by postulating the identity of some events, it must quickly come also to postulating the identity of two objects which one might have thought were different. Mind and brain are inevitably the same object, and one might as well call them by the neutral term 'person'.

The second doctrine which, working together with the Identity theory, allows us to avoid drifting into an extravagant theory of mind is the doctrine that whereas causal laws are intensional relations between *descriptions* of events, causation itself is an extensional relation, holding between brute events. This principle so stated applies, of course, to event causality. When we extend it into agent causality we must amend it to this: causation is an extensional relation between brute objects and brute events, and a causal account (since here we have no laws) is a relation between a description of an object and a description of an event.

Let me explain the relevance of this point with some care. The idea which was driving us to think we might have to postulate a special sort of mental object like a Mind or Self was this. For the mental description over the neural one of the decision to be hegemonic it must be caused (or anyway explained) in a stronger way than the neural description is. We have managed to get that mental description caused by a piece of agent causality. But – so the thinking runs – the only sort of object that can cause a mental event is surely a mental object; therefore we must have a mental object.

As soon as we make this line of thought explicit in this way, the error in it becomes apparent. One doesn't need a mental object to cause a mental event: the causal relation holds between brute objects and brute events – the descriptions of those items come into play when we try to write up laws or accounts covering the causal relation. As far as the causal relation itself goes, then, we need no mental object. Neither do we need a mental object to enter into the causal account: what we need there is a mental description of an object. But even that is a bit hasty: what we need for the causal account is not exactly a mental description of the object (whatever that would be) but rather a description of the object which will enter into a causal account that renders the decision under its mental description. A description of an event is hegemonic when it is more strongly rendered in explanation than other descriptions of the same event; there is no strict requirement set about the nature of the terms used in the explanation: all that is required is that it be a good explanation.

Altogether then, we do not need a mental object to enter into the causal relation; we certainly do not need a mental object to enter into the causal account; we do not even need a mental description of the object to enter into the causal account. All we need is to be able to describe the object in such a way that that description will enter into a satisfactory causal account (or, more generally, explanation) that will strongly render the mental description of the event and make it hegemonic.

I suggest that in the case of the man deciding to divorce his wife such a satisfactory description of the object is 'he', 'the man' or 'the person'; the explanation which renders the decision under the mental description ' "divorce wife" appears as *decisum*' is simply 'he made "divorce wife" appear as *decisum*': the mental description of the event is hegemonic because while 'he made "divorce wife" appear as *decisum*' is a satisfactory, if stilted, explanation of the event, 'he made such and such a neuron fire' is not: in fact it makes no sense at all.

It may seem that we have had a terrible labour for so humble a result. But it is indeed encouraging that these various metaphysical doctrines collude so neatly to show not only that we have no need to postulate extravagant mental objects but even that we properly cannot do so. The theory of agent causality by which we have solved the libertarian's third large problem entails no revision to the theory of mind that we have been employing all along.

(v) The conditions of freedom of decision

The libertarian began with the view that one of the necessary conditions for the freedom of a decision was that it not be necessitated by antecedent causes; he could not at first say, though, what the sufficient conditions might be. We have now developed an account of them, however, and perhaps it would be well to conclude by trying to state them succinctly.

The conditions are (a) that the decision not be an event which is causally necessitated by an antecedent event, and (b) that the decision be caused by the agent whose decision it is. (In every reference to agent causality I mean what I have called radical agent causality.) It might seem puzzling that this statement of the conditions makes no mention of the question which we earlier mentioned as the test for liberty of in-difference: could he have decided otherwise?

If for the purpose of argument we suppose that there is this third condition, (c), that he could have decided otherwise, then we might glimpse the possibility of a pleasing result. The third condition entails the first: 'he could have decided otherwise' entails that it was naturally possible that he decide otherwise; might we want to say that the second condition entails the third, that 'the decision was caused by the agent' entails that 'he could have decided otherwise'? If so, that is if (c) → (a) and (b) → (c), then the second condition, agent causality, implies the other two. And if that is so then we have a very succinct statement indeed of the sufficient conditions for freedom: there is just one con-dition, that the decision be caused by the agent in an instance of radical agent causality. The three conditions have telescoped to one.

The hope for such a clean result may seem too sanguine, however. What reason is there to believe that 'the decision was caused by the agent' entails 'the agent could have decided otherwise', that (b) entails (c)? Well, here is an argument that the entailment holds. In the course of this present chapter we offered the proposal that the idea of radical agent causality was equivalent to the idea of activity, that an agent is active in all and only those events which he causes. Substituting by this equivalence, then, we must show that if an agent is active in a decision then he could have decided otherwise, in order to demonstrate that (b) → (c). And this proposition can perhaps be demonstrated contra-positively. If we show, that is, that if he could not have decided other-wise he was passive in the decision, then because being passive entails not being active, it will follow that if he was active he could have decided otherwise.

Some thought-experiments might perhaps convince us of the truth of the counterfactual entailment, 'if he could not have decided otherwise he was passive to the decision'. The difficulty is that it is hard to find cases where we are prepared to say that an agent really could not have decided otherwise, and where the 'can' is not at all 'iffy'. But suppose an agent is hypnotized and instructed that later that week he shall consider whether or not to divorce his wife, and shall decide in favour of divorce. Suppose too that hypnosis is such that it entails that the agent cannot but obey the hypnotist's instructions. The end of the week comes and the agent decides to divorce his wife. Are we not inclined here to say that in his decision the agent is a victim, a patient? Are we not inclined to say that, for all its appearance to the contrary, this decision was something which happened to the agent, which was imposed upon him? Are we not inclined to say that he was passive, not just nominally but really, in the decision?

Or if one believed that some sort of universal determinism by the subconscious, in the manner of Freud, were true, would one not regard an agent in each of his decisions as a victim of his subconscious, as passive – no matter how active he seemed to himself to be? Here too, then, his inability to have decided otherwise entails his passivity in the decision.

It follows from all of this, in the way that I set out above, that the second condition for freedom entails the third, and altogether therefore that the second condition on its own, radical agent causality, amounts to a complete and succinct statement of the sufficient conditions of freedom. A decision is free if and only if it is caused by the agent whose decision it is, in an instance of radical agent causality.

We must bear in mind, though, that what we have here been elaborating is the conditions for freedom of *decision*. When we turn to freedom of action the conditions become more complicated in an extremely interesting way.

VII

PARALIPOMENA

The principal task of this book is now completed. We have guided the libertarian through the three difficulties which seemed at the outset to threaten the tenability, and even the coherence, of his theory. First, we have found that nothing very incredible would have to be true in neurology for the brain state to lack sufficient causal conditions for being what it is, and indeed to lack such conditions most of the time. Second, by mere reflection on our intuitions about parallel cases, we have shown that the brain state may often depend upon the mind state: that was our theory of hegemony. Third, we have battled through to a solution for the really deep problem about the difference between randomness and freedom: we have argued that the libertarian's solution to that problem requires some conceptual revision at a deep level; we have offered such a revision and filled in a certain amount of the terrain surrounding it. Altogether, then, we have shown the libertarian position to be both coherent and possible; since the most forceful seeming attacks on libertarianism argue that it is not coherent, we have given it the sort of defence that it most urgently needed.

The kernel of libertarianism, then, is sound. There are still a number of problems on the periphery, however, and it is to these that we now turn. The first of these is that it is not normally for our decisions but for our actions that we are held responsible: the libertarian must offer us some story about how his account of freedom can be extended to cover actions as well.

The second peripheral problem has to do with freedom of will and strength of will. These two ideas, surprisingly, are rarely discussed together. If our will is free, absolutely free, what need has it of strength?

How is strength of will to be construed?

The third and final peripheral problem has to do with scalar freedom. We have argued that freedom is an absolute, all-or-none affair; but how is that to be squared with our common way of speaking in which freedom is often said to be a matter of degree? And how are we to accommodate the clear implication of libertarian theory that animals and children enjoy free will? For that implication runs contrary to generally received views.

(i) Freedom of action

Many of our acts are not preceded by anything that could be called a decision to do those acts – and yet we suppose those acts, or at least many of them, free with the liberty of indifference: we could have done otherwise. But can liberty of indifference belong to anything but decisions? Can there be *arbitrium* in a case where there are not alternatives being considered?

The libertarian can answer that decision is not the only case of liberty of indifference, though it is a special and conspicuous case of it. In a decision we are considering, say, whether to do x or to do y. We decide to do x. The burden of 'we could have decided otherwise' is then naturally 'we could have decided to do y'. But let us take the case of an act which is not preceded by a decision – ordering a firing squad, for example. I say 'fire!'. That is an act which may well not have been preceded by a decision, and yet we should want to say that I could have done otherwise. But could I really, if I weren't considering alternatives, if I weren't deciding? Surely a condition of having been able to do otherwise is having had some alternative course of action in mind.

What the libertarian must say here is that the burden of 'I could have done otherwise' is at least 'I could have refrained from saying "fire!" ', and that the possibility of refraining is a possibility of which any agent is always conscious. Decision is typically a case of choosing between x and y; but free agents are always engaged in choosing at least between x and not-x. Of course the libertarian will insist that if this act, saying 'fire!' is free with the liberty of indifference, it must be an act which is caused by the agent in an instance of radical agent causality, and that entails that it is not naturally necessary. This in turn will be possible if the neural indeterminacy we imagine is more frequent an occurrence than decisions are; there seems to be no difficulty about that. Indeed there seems no real difficulty about saying that the neural state of a man is more or less permanently in a state of indeterminacy; this would match the point that an agent is more or less permanently conscious of the possibility of refraining from what he is doing.

Of course, an agent can be the cause, in the radical sense, of only a limited range of acts: the mental changes or bodily motions which traditionally are said to be subject to the will; finally these are what the agent can do or refrain from doing. Roughly, these are identical with the 'basic acts' of modern action theory. By allowing that an agent can (in the radical sense) cause these acts directly, the libertarian can escape the old embarrassment of having to postulate acts of will before each free act - acts of will which simply do not seem always to be there. What confers freedom on an act is not another act, an act of will, but rather its being caused by an agent in radical agent causality.

There does however arise an intriguing problem over freedom of action which did not arise over freedom of decision. At the end of chapter VI we gave as a full statement of the sufficient conditions for free decision, the following:

(a) the decision was not causally necessitated by preceding events;

(b) the decision was caused by the agent, in the radical sense;

(c) the agent could have decided otherwise.

We saw that these three telescoped to one, (b), because (b) entails (c) and (c) entails (a). In the case of freedom of action, though, this telescoping does not succeed. There the conditions would be

(d) the action was not causally necessitated by preceding events;

(e) the act was caused by the agent, in the radical sense;

(f) the agent could have done otherwise.

The telescoping does not succeed because of an ambiguity in the third condition. This ambiguity will take some careful exposition.

Although philosophers have largely discarded the idea of the will, physiologists and anatomists have a ready sense of it. A glance at any anatomy textbook shows that one of the points that is mentioned in the description of any muscle or organ in the body is whether or not its motion is 'subject to the will'. The beating of the heart is not subject to the will, but the opening and closing of the epiglottis is. What the physiology of this being 'subject to the will' amounts to is generally described, darkly, as being controlled by 'higher centres of the brain', and there is no doubt that that much, at least, is true. We don't yet know the mechanism of this with any exactness, but we do know that the impulse for a voluntary movement originates in the skull, whereas the motor impulse for a non-voluntary movement originates much lower down. I don't mean to suggest that the distinction between voluntary and non-voluntary movements (or muscles) is absolutely clean, but at least we have the general picture.

Now let us suppose that we come to have a better understanding of the physiology of 'being subject to the will'; let us suppose, undoubtedly making matters simpler than they are, that the motor nervous impulses for all and only those movements that are voluntary begin in region W of the brain. The supposition is simplistic in form but not in principle: there must be some neurological correlate of the voluntariness of voluntary motor activity.

On this model so far the libertarian and the neurophysiological determinist are in perfect agreement. The determinist thinks that that is enough to say to characterize the voluntary. The libertarian thinks that something more must be said. He maintains that the initial neural event in W that starts off a motor message which ultimately leads to voluntary motor activity must be an event that is not caused by preceding events; he maintains that it is caused by the agent albeit under another description. But it is of course only the initial event in that neural train that is so caused: the other events in the train are caused each by the one before. Now strictly speaking, therefore, all that the agent does is to start the train going; once it is going, it takes care of itself. And here is where the difficulty begins.

In various abnormal circumstances a normally voluntary muscle can be stimulated by an impulse that bypasses W. Perhaps the most dramatic illustration of this was the work by Penfield on the motor region of the cortex. When a point on the cortex was stimulated with an electrode the patient's arm would go up or he would emit a cry, and be utterly surprised by having done so.[1] The situation is illustrated in Figure 7.1.

The agent can directly make the first neural event at W happen, or stop it from happening, but he more or less relies on things to be normal in the nervous system for that first neural event to have its customary ultimate effect, the lifting of the finger. If however things are not normal and the neural tract from W to the finger is inhibited at P by impulses derived from artificial stimulation on the cortex at Q by an electrode, he can make the initial event at W happen, but that event will not have its usual train of effects. Or if the agent voluntarily refrains from causing the trigger event at W, then the stimulation derived from the electrode at Q might enter the pathway at P and cause

the finger to move. It is with this sort of model that we would portray
what was happening in the Penfield experiments.

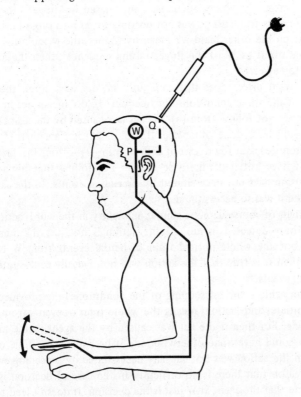

Figure 7.1 'He could have not raised his finger'

Now I said that all this produced an ambiguity in the third condition for freedom of action, that the agent could have done otherwise.
The ambiguity is whether we mean by 'could have done otherwise'
could have accomplished a doing otherwise, or could have initiated a
doing otherwise.

Let me give an example to make the difficulty clear. Suppose that at
t_1 the electrode at Q started a stimulatory impulse which reaches P at
t_3; and suppose that at t_2, in ignorance of the electrode, the agent –
what would one say? – wills, tries, to lift his finger; that is, he causes
the neural event at W which would in the normal course of events lead
quickly to the lifting of his finger. That impulse reaches P at t_3 and,

overdeterminedly, the finger lifts. Was this raising of the finger a free act? Could he have done otherwise?

If we mean by 'done otherwise' 'not raised his finger', then the answer appears to be no. It was not possible for him to avoid raising his finger. If on the other hand we mean by 'done otherwise' 'not caused the initial event at W of the finger raising sequence', then the answer appears to be yes.

Now, what effect does this ambiguity in 'do' have upon the telescoping of the three conditions for freedom? If 'do' means accomplish, then it does not follow from (e) 'the act was caused by the agent', that (f) he could have done otherwise. If 'do' means initiate then (f) does follow from (e), but (d) does not follow from (f) – from the fact that he could have initiated otherwise it does not follow that his action, finger raising, was not necessitated by preceding events. In the case we considered it was so necessitated.

We might of course detect a similar ambiguity in the word 'action' to that we have detected in 'do'. If the action is not just the muscular activity but the whole normal train of neural events from W to the muscle, then it is true that the action was not causally necessitated by preceding events.

We can achieve the telescoping of the conditions only provided 'do' means initiate, and 'action' means the whole train of events from W to the muscle. For then if the act was caused by the agent in the radical sense he could have done otherwise, and if he could have done otherwise then the action was not causally necessitated by preceding events.

The reason that there is no difficulty in the case of decision is that we assume that the event at W just *is* the decision: it doesn't lead to the decision, but just *is* it. If that assumption is false, of course, all that we have said about the problems of telescoping the conditions for free actions will apply also to decisions.

What all of this shows is that the psychological unit the 'basic act' is not the same unit as what the libertarian neurologist would think of as the event that is directly and freely caused by the agent. The latter is only a portion of the former. No doubt the rather stark and extreme difference between them that emerges from this account can be softened somewhat by a more complex neurophysiological model (a model say, that requires a series of W-events to keep the impulse going down the track), but nonetheless the difference remains.

The other point that it leads me to suggest is that perhaps talk of volitions preceding all voluntary acts is not so hopelessly wrong-headed as it is fashionable to assume. On the model we have presented an event at W is the condition of a voluntary act; it is like a volition. The event at W has, all on its own, no correlate in consciousness – volitions are not mental events. The smallest psychological unit here is the act of raising the finger. Moreover the W-event is not an *act* that precedes the *act* of raising the finger; rather it is the first event in the act of raising the finger. We have an obscure sense of how our nervous system functions: the postulation of volitions may reflect that sense.

(ii) Freedom of will and strength of will

The libertarian must extend his theory to answer some problems that are posed for it by the collection of phenomena which we organize with our concepts of strength and weakness of will. The first of them can be put in this way. Why are people so predictable if they really are free in their decisions and actions? If what underlies free decisions and actions is a piece of neural indeterminacy, surely there should be utter unpredictability of people's actions. Why, for a given person, is one decision so much more likely than another? When a fat lady who has told herself that she must diet is placed before a chocolate éclair, why is it so much more likely that she will eat it than that she will not? The case of akrasia is, I suppose, just a very dramatic case of the strength of habit in our lives. Surely if the libertarian is right in his understanding of freedom, habits should be easy to break – indeed they should never even be formed – and all wills should be equally, perfectly, strong.

One can put this problem as a problem about probabilities. If we really take seriously the libertarian's idea that the indeterminism which occurs in the nervous system is real, non-epistemic, indeterminism, it seems as though the probability that an event occur or not cannot be weighted. That is, if it is possible that the neuron fire or that it not fire, then the real probability of each alternative must be 0.5. But how then can it be so much more likely that the lady will eat the éclair than that she will not?

Before I try to answer this question let me be very clear as to why I say that the probability of the cell's firing or not cannot be a weighted probability. (I mean, of course, a real, non-epistemic probability, just as I mean a real, non-epistemic indeterminism.) It might be thought that this is wrong, that the probability can be weighted, on the following grounds. Let us say that the alternatives are that the cell fire or that it not fire. Suppose we use the indeterministic self-emptying tank model of chapter IV, and say that the cell cannot fire with fewer than 97 units of charge built up, and that it must fire if 108 units of charge build up, but that between these two figures there are sufficient conditions neither for its firing nor for its not firing. Suppose further that right now there are 106 units of charge built up. One might feel very inclined to say that, with 106 units of charge, it is far more probable that it fire than that it not fire.

But this would be wrong. The thought that leads to this mistake is that although having 106 units is not sufficient conditions for firing, it is much closer to sufficient conditions for firing than it is to sufficient conditions for not firing: it is much easier to make up the two than to subtract the ten. That is, from the present position it is easier to achieve sufficient conditions for firing than to achieve sufficient conditions for not firing.

But that is to think that to fire we have to get sufficient conditions for firing, or not to fire we have to get sufficient conditions for not firing. And that is the mistake. A causally undetermined state or event is precisely one that obtains or occurs without having sufficient conditions for obtaining or occurring. It is not a question of making up sufficient conditions, but of the event's happening without sufficient conditions. If the indeterminism is real, therefore, the real probability of each of the alternatives is one divided by the number of alternatives. In the case of the neural correlates of eating or not eating the éclair, the probability of each must be 0.5. The question stands: why is it so much more probable that she will decide to eat the éclair than that she will not?

The answer I think is that the greater probability that she will eat the éclair than that she will not attaches not to the neural descriptions of these events but to their mental descriptions. Probability, even real non-epistemic probability, attaches to descriptions of events, not to brute events themselves. We must explain how this works.

I think it is rather simple. The characteristic of weak-willed people is that the neural correlate of resisting this or that temptation carries along with it the neural correlate of tension and discomfort – call it 'pain'. Resisting temptation induces pain in the weak-willed: that is what it is like to be weak-willed. People dislike pain and they are likely (but not forced) to take avoiding action. This likelihood or probability does not belong to the neural descriptions, however, but purely to the mental descriptions.

Incidentally, the mistake of supposing that the unbalanced

probability at the mental level must arise from or be reducible to an un-balanced probability at the neural level stems in part from a misconception which dogs and confounds the science of neuropsychology at every turn. It is the misconception that the structure of felt experience must be mirrored in the structure of neurophysiology. It is under the influence of this deep-lying misconception that we have come to suppose that since we feel pressure from the desire to eat chocolate, it must be the case that the neural correlate of the desire is exerting pressure, electrical pressure it would have to be, on the cells which would start off the act of eating – W, perhaps. Because the felt experience has the character of pressure which builds slowly to the breaking point, we suppose that the neural correlate of the desire must be sending stimulus ever faster and more furiously towards W, until in the end W is overcome and sends its signal to initiate the devouring of the éclair.

But it need not be so at all, of course: there need be no such mirroring of structure. The nervous system may be perfectly serene; there need not be any neural battering at the gates of W. It can simply send more frequent pain messages to the cortex, and that results in intense pain. The intensity may all be in the pain, that is, at the mental level. And pain influences the agent to take avoiding action. The nervous system need not be trying to take avoiding action itself: it may simply be influencing the agent to do so. If the nervous system could take avoiding action itself, pain would be biologically useless. Indeed it has often been remarked that pain seems to be biologically useless; but the libertarian can make sense of it. On his view the nervous system, in so far as it is indeterministic for voluntary action, cannot directly cause avoiding action to be taken. All it can do is influence the agent to do so by causing him pain. To the mind it seems as though there is something like a causal link between the pain and the behaviour; this causal link need not, however, obtain between the neural correlates of the pain and those which initiate the behaviour.

Another way of putting this point about structural incongruity is this. It is much harder for the lady not to eat the éclair than it is for her to eat it. We are not to suppose that it is much harder for W not to send its trigger signal than it is to send it. There need not be parallelism of structure in these functions.

We must now work up an explanation along libertarian lines of what it is for different people to have different strength of will. The very language of weakness and strength of will, of course, suggests determinism. It suggests a model of something like strings of different tensile strength. Some people will endure only a small amount of tension (temptation) before they snap (yield): others will endure much more; in either case they yield exactly when the tensile strength of the will is overcome.

But of course this will not do at all for the libertarian. He will doubtless want to allow different absolute breaking points to different people, but also a large stretch of indeterminacy beneath those absolute breaking points where the agent may or may not give in.

In fact, given what we have said about the lack of structural parallelism between the felt experience and the neurology, there need be no absolute breaking point at all except in cases of sheer muscular endurance. What I mean is that it may be the case that at no point does it become physically necessary that a spy who is being tortured reveal his secret: he may break in the sense of fainting or even dying with pain, but the pain will not cause him to speak. On the other hand a climber who has lost his foothold and is hanging by a rope may reach the point where it is physically necessary that he lose his grip. But there it is not the pain that causes him to let go but the fact that the sheer physical energy of his body that is required to tense the fingers gives out. Undoubtedly he would faint from pain before that point, however, and in that rather indirect sense the pain would cause him to lose his grip: it would not, however, cause him to loose his grip.

The theory towards which all of this drives us is the following. Pain is never a cause of actions, but only an influence towards them. (Remember here that 'pain' is being used to stand for the range of mentalistic terms of discomfort: distress, tension, grief, etc.) And influences are radically different from causes. We might propose the following analogy. Think of the control room of a generating station. Imagine that what happens when one of the generators becomes too hot is that a red light comes on and a buzzer sounds. The operator, seeing those signals, presses a lever which causes oil to be pumped into the turbine so that it runs with less friction. At the level of the electrical mechanism there is no causal link at all between the monitory trouble signal and the motor signal which causes the oil to flow. The link between those two happens outside the electrical mechanism – the control panel. In a similar way, the libertarian proposes that there is no causal link in the neurophysiology between the pain correlates and the first firing of the motor neurons that are concerned. The link, which is not a causal link but a link of influence, occurs at the level of the mind.

I have spoken darkly of influences as different from causes; it is important to be clear about this idea. We can begin by extending the analogy. The flash of the monitory light and the sound of the buzzer do not cause the man to press the lever; they do not amount to sufficient conditions for his doing so. Perhaps in a very extreme case with a very bright light and a jarring buzzer they would ultimately cause him to press the lever, that is, make it inevitable that he should, but not in the normal case. (Of course this is only an analogy and its weakness is that if one is disposed to see the human being as a deterministic mechanism one will be unimpressed by it: one will believe that other things being equal the bell and light do cause the man to press the lever. It is one of the difficulties of the philosophy of mind that there are no universally persuasive analogies for the mind; the only analogy for the mind is the mind, and the view of the mind to which one is predisposed will determine the strength of the analogy.) Annoyance with the light and the buzzer is an influence on the operator's behaviour, not a cause of it.

This distinction I understand as a deep one; it can perhaps be explained in the following way. It often seems irresistible to regard the mind as a sort of field in which there operate vector forces - the various feelings, desires and motives to which we are subject (Figure 7.2). It is a

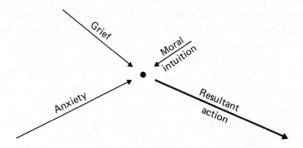

Figure 7.2 The vector-field model of the mind

part of this picture that the action we take in a situation is the simple resultant of the various forces which act upon us. This is a model which gives rise to psychological determinism. Now in this model there would perhaps be some place for a distinction between influence and cause, but it would be a shallow one. The cause of one's act might be identified as the strongest of the forces acting upon one (or the most unusual, or the one that is for some reason most interesting . . .), and then all the others would just be influences. But there would be no deep difference between the two – just a difference of strength or odd-ness: fundamentally all the forces work in the same way and on the same object.

The model that the libertarian will want to propose is such that the forces of pain do not ever meet in a point, and so there is no resultant formed (Figure 7.3).

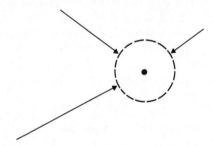

Figure 7.3 The libertarian's model of the mind

The agent is indeed aware of them and they influence him, but they do not cause him to do anything. Rather he causes himself to do what he does, and he does so in the knowledge of the forces that act upon him. The force of pain, we might say, takes its special nature as an influence from the nature of the object upon which it works. But since a mental force cannot ever work on anything except an indeterministic agent, the transference seems justified.

If one were to object to all of this that it does not reflect the experience of temptation, torture, or duress, I shall reply that it does, if only one looks closely enough at that experience. We are always aware that we could hold out, just a bit longer, and afterwards often remorseful that we did not.

On those extreme occasions, if any, when an agent is caused by pain (again, I use the word generically) to do something or other, the libertarian will want to say that the causation took place not at the mental level – for there there can only be influence – but at the neural.

Furthermore he will be attracted by the hypothesis that the neural train which in such an extreme case causes the agent to, say, lift his finger (Figure 7.1) is one which does not go through W, so that the action would not count as voluntary even in the anatomist's sense. This of course is an empirical hypothesis which would need to be tested; in as much as it would require subjects to try the outermost reaches of their endurance it probably never will be tested. If this hypothesis is right, then even such extreme pain cannot force one to *do* an *act,* for doing and action require neural activity at W: the dangling climber can be made to lose his grip, but not to loose it.[2]

Very rarely, however, is life conducted at such extremes of influence that the language of causation becomes appropriate. Much more often there is no real sense in which we have to do the things we are tempted to do. The difference here between weakness and strength of will is that the strong-willed person feels little pain in a temptation whereas the weak-willed person feels much. But that answer, of course, is not to be extended into a deterministic model in which all persons can endure the same amount of pain, the strong having the good fortune to reach that critical point rarely and the weak the bad fortune to reach it often. What the libertarian must say, rather, is that, yes, the fat lady could resist the temptation to eat the éclair, just as the dispositionally thin lady can; but to do this would be far more taxing to the fat lady than to the thin.

(iii) Scalar freedom and responsibility

The libertarian theory we have presented is triply uncompromising. In the first place it regards freedom as absolute, all-or-none: the answer to the question 'could he have done otherwise?' is 'yes' or 'no'. The 'could' involved is the could of physical possibility – and that is something which either you have or you don't. Second, the theory is uncompromising in that it insists that we have such absolute freedom in virtually all of our acts, at every minute of the day. The only cases where we might lack this freedom would be such cases as occur in the lives of very few persons, when they are fiercely pressed to the outermost limits of their endurance by some or other form of pain. Third, this uncompromising freedom is possessed not only by adult persons but is inflicted on children and animals as well.

All of this, of course, flies in the face of many of our ways of talking about freedom. That just in itself is not a direct embarrassment to the libertarian, for it was clear from the outset that he wanted to investigate only one of the several ways of talking about it. But the metaphysics that he has developed to make sense of that one way of talking ('liberty of indifference') has in a sense run away with the linguistic phenomena it was designed to explain. That is, having wanted to take seriously the idiom 'I could have done otherwise' he has worked up a metaphysics to ground it in; but that metaphysics, taken seriously, turns out to imply that we could have done otherwise in all sorts of cases where we should be slow to say that we could have. The libertarian must sort this out. He must say how it arises that language and practice become less uncompromising than his metaphysics would direct.

The first issue then is whether freedom is absolute, for we often speak as though it is scalar. We say for example that so and so was not a totally free agent at the time – he had been threatened by a gangster. How are we to understand this scalar way of speaking? I sketched out an answer to this in chapter I (pp. 7–9): talk of diminished freedom arises by tincture with talk of responsibility. We have a tendency to think that in so far as we are free we are responsible and that in so far as we are responsible we are free. But we do find ourselves talking of diminished responsibility[3] in many cases, and so we are led to talk of diminished freedom. I construed 'diminished responsibility' as itself a *façon de parler,* though; properly, responsibility is an absolute notion: either one is responsible or one is not; either one can be held to account or one cannot. 'Diminished responsibility' is really absolute responsibility for a crime of diminished heinousness, and the heinousness of a crime is diminished by the pain or duress or other such conditions under which one commits it.

We are now in a position to fill out this account and to make it more subtle. It is true that we sometimes have the idea that responsibility is absolute, and at other times that it is scalar. We ought I think to distinguish two senses of the term, a metaphysical sense and a moral sense. In the metaphysical sense one was responsible or one was not – it is, shall we say, a matter of fact. One was responsible if one could have done otherwise, and if not, not. In its moral sense, 'he is responsible for x' means 'he ought to be held responsible for x', and this moral sense of being responsible admits degree. Roughly, the more difficult it would have been not to have done as one did, the less one is held responsible for it.

The libertarian then, will explain this ambiguity in this way. Although it is always, or very nearly always, possible to do otherwise, sometimes doing otherwise would involve so much pain or difficulty that society does not ask or expect it of us. What is expected of us in this line is a moral matter, a matter of the *mores* of a given group, and the expectations can vary from group to group and even within a group from time to time. In our society, we permit cannibalism only in the direst circumstances; in others much less distress is needed to justify it. The Greeks apparently thought that no form of duress or pain could ever justify matricide;[4] we count it along with other forms of murder, and make no special rules for it. Even more interesting is that this distinction will allow us to explain such puzzling facts as that French law recognizes the *crime passionnel* where English common law does not: French law does not expect a person to withstand the provocation of discovering the infidelity of a spouse; common law does lay that expectation on us.

The question, then, as to how much tension we should be expected to endure before giving in to temptation is a moral matter; that expectations in this regard should vary dramatically from place to place and from time to time is entirely consistent with its being the case that strictly, metaphysically, one could have done much better than the expectations, one could have held out until unconsciousness or sleep came on. Indeed any theory about these matters must allow for the phenomenon that one person may at the same time be subject to different expectations from different groups to which he belongs. Society may for one reason or another excuse the parent who was negligent of his children, but the parent may not excuse himself; society may not hold him responsible, but he may hold himself responsible. Again society may not hold a spy responsible for having broken under torture, but the secret service that employs him may do so.

So far, then, we can summarize what we have said about scalar freedom in this way. Saying that freedom is scalar is compatible with saying that it is absolute because its absoluteness has to do only with the presence or absence of causes of the action, whereas scalarity has to do with the presence or absence of influences; and influences, we recall, are radically different from causes. No sum of influences amounts to a cause.

The second way in which the libertarian theory seemed too uncompromising was that it held that we have absolute freedom all the time. C. A. Campbell distinguished what he called expressive willing from creative willing, and held that the former goes on most of the time and the latter in isolated moments of moral effort.[5] In expressive willing our acts simply reflect our already formed characters; in creative willing we transcend those characters. Something like this is descriptively correct for most of us, no doubt, but we must be careful how we understand it. It is not that most of the time we are unable to be creative in that sense and so can only be expressive; it is rather that although we at all times could be exercising the creative will we usually are not. Another way of putting the same point, a way which exhibits the ambiguity in this way of speaking, is to say that we are always and inevitably exercising the creative will, but most of the time we are using it to entrench what we already have by way of character, and at other times we use it to extend or reform what we have. We must not think that because we are not at every moment reforming ourselves we could not be doing so: it is always, or very nearly always, possible to do otherwise.

Here again we have uncompromising metaphysics and compromising morals, and there is nothing wrong with that. Only a fanatic would expect us to be engaged in reforming ourselves the whole time, and here the fanatic is guilty not of a metaphysical error but of a moral one. It is possible but not normally expected of us that we be constantly actively aware of the habits we are building or losing. By making this distinction between the metaphysical and the moral senses of responsibility and so of freedom the libertarian, incidentally, leaves room for an important kind of moral reform: he leaves room for any programme, shall we say, of raising expectations as to endurance or self-watchfulness, whether it be in a Church or a therapy group, or even in a society at large. Such a reform entails but a change in *mores*; it does not stub its toe on metaphysical bedrock.

Finally, and most interestingly, the libertarian faces the problem that his theory ascribes freedom to children and madmen and animals, and this certainly seems very odd. Here again I think that the problem arises because the notion of freedom is coloured by that of responsibility. The notion of freedom defended by the libertarian is equivalent to metaphysical responsibility but not to moral. We do not hold animals and children responsible and so we think they cannot be free. This inference plays on the ambiguity between metaphysical and moral responsibility: we do not hold animals and children responsible, that is, they are not morally responsible, and we do not even raise the question of their metaphysical responsibility.

Here, however, the reason that we do not hold them responsible is different from what it was in the previous cases. It is not that they act under duress, distress, or pain. It is rather that moral responsibility is a relative notion: one is morally responsible to a community; one is held responsible by a community; one must be a member of that community to be held responsible by it to it. Animals, children, and madmen are not members of the community whose institutions we philosophers analyse. Because they should not be held responsible to our community, we suppose they should not be held responsible at all. But that is a mistake. Clear and much-studied institutions of responsibility exist among children as soon as they are old enough to have a community among themselves. There seems no clear reason why fairly advanced animals should not be said to have the rudiments of something similar. And if a group of madmen recognized a reasonably complex public world in common there seems no reason why they should not have the institution of responsibility among themselves. The fact that these various groups should not be held responsible by us does not entail that they should not be held morally responsible. And if they should be held morally responsible (by anyone) that is because they are metaphysically responsible: they were free in doing whatever they did – they could have done otherwise.

The libertarian must say that when we deny freedom to animals, or children, or madmen, we are confusing a number of things. We are confusing being morally responsible with being morally responsible to our community, in the first place, and in the second we are confusing being morally responsible with being metaphysically responsible, and so with being free. If the confusions were all sorted out, the pattern of the argument behind the idea that animals, madmen and children are not free would be:

1. Hieronimo's mad againe;
∴ 2. Hieronimo is not morally responsible to our community;
∴ 3. Hieronimo is not morally responsible;
∴ 4. Hieronimo is not metaphysically responsible;
∴ 5. Hieronimo is not free.

But steps 2–3 and 3–4 are invalid, as we have seen.

The question of moral responsibility – let alone that of responsibility to any particular moral community – enters into very few of our acts. All, or nearly all, those acts, however, are free: it is always, or nearly always, possible to do otherwise.

NOTES

I INTRODUCTION

1 Pamela Huby argues that the first philosopher really to see the problem of freedom and determinism was Epicurus; Democritus and later Aristotle had all the equipment to appreciate the problem, and were on the verge of appreciating it, but did not quite succeed in doing so. 'The first discovery of the freewill problem', *Philosophy*, 42 (1967), pp. 353–62.

2 It is something of a mystery to me that, whereas there is a long Latin tradition going back through Valla and Augustine to Boethius which knows our problem as the problem of 'liberum arbitrium', the standard English phrase should be so theory-encrusted as 'free will'. That the latter could be intended simply as a rendering of the former is shown in Chaucer's translation of Boethius' *de Consolatione philosophiae*, where 'liberum arbitrium' is rendered 'free wyl'; however it is often intended as more than that, and to avoid confusion I shall use the phrase 'free decision'.

3 In fact, for Campbell, the class of free actions seems to be even more restricted than this: *In Defence of Free Will*, London, Allen & Unwin, 1967, p. 49 & p. 57. On p. 49 Campbell seems to argue that freedom of choice occurs only in a case of 'effortful willing', and on p. 57 he admits that effortful willing is absent from 99 per cent of our choices. But it is surely wrong to think that a choice is free only when it is for the more difficult alternative. This view arises by tincture with liberty of spontaneity, which I shall shortly discuss.

4 There is another advantage for the libertarian in discussing freedom of decision: it places him in an initially stronger position for doing battle with compatibilism, should he wish to enter that field. 'He could have decided otherwise' is not nearly so susceptible to the suggestion that it understands the enfeebling protasis 'if he had

142

wanted to' as is its counterpart 'he could have done otherwise'.

5 Thus Bergson, for whom, in the famous remark, 'the free action drops from the self like an over ripe fruit': *Time and Free Will,* London, Allen & Unwin, 1950, p. 176. So too Wiggins, though less covertly, in 'Towards a reasonable libertarianism', in *Essays on Freedom of Action,* Ted Honderich (ed.), London, Routledge & Kegan Paul, 1973.

6 It is preeminently to liberty of spontaneity that Austin's famous suggestion that 'free' is to be analysed as 'free from —' belongs.

7 I use the convention that capital letters, M, N etc. denote event kinds, and lower case letters, m, n etc. denote events.

8 There is perhaps a weakness in this example stemming from the fact that usually the ingestion of aspirin has a mental correlate, viz., seeming to take aspirin: there seems to be a nearly lawlike connection between the mental event kinds, seeming to take aspirin, and losing a headache. The weakness of the example can be overcome by supposing that someone has surreptitiously dissolved the aspirin in the onion soup.

9 The notion of 'strength of explanation' which is here offered informally and loosely is presented in a more elaborate and more formal fashion in chapter V, section (i).

10 Implication rather than equivalence here because the Correlation thesis we are considering will allow a mental kind–neural kind correlation of one-to-one or one-to-many, but not of many-to-one or many-to-many. This is explained in chapter III.

11 And he will offer, or at any rate think offerable, such a law for any given kind of decision. The formal statement of his claim will be: where P ranges over kinds of total physical state in and surrounding a man at a time, D ranges over kinds of decisions of that man, x, y, a, b over events or states,

$$(x) \left[(Dx \text{ at } t') \rightarrow \{ (\exists y) \, (Py \text{ at } t) \land [(a) \, (Pa \text{ at } t) \rightarrow (\exists b)(Db \text{ at } t')] \} \right]$$

12 Wiggins formulated the argument which gives rise to this problem in 'Towards a reasonable libertarianism' (see n. 5 above). The formulation which I offer differs from his in a number of respects:

(i) He was concerned only with physiological determinism and so he does not need step (2), and thus, also, his R symbolizes not a mental event but a bodily movement. His formulation is:

(I) inevitable at t' (C at $t \rightarrow$ R at t')

(II) C at t

∴ (III) inevitable at t' (R at t')

(ii) His modal operator 'inevitable at t'' seems to claim in (I) much less than the determinist wants to claim, and consequently in (III) to derive a conclusion which the fiercest anti-determinist would acquiesce in. For surely the anti-determinist won't mind if it is established that it is inevitable at (or after) t' that R at t'; what he wants to resist is the conclusion that R at t' is inevitable *before* t'. Indeed, 'inevitable at t' (R at t')' is true by the self-necessity of present and past fact, if R at t'. And if we allow that sort of necessity – as we should and as Wiggins ultimately does – then if we are given that R at t' (i.e. if this argument is a post factum account of, rather that a prediction of an action) then (I) can be shown to be trivially true, being implied simply by R at t':

1. R at t' (given)
∴ 2. inevitable at t' (R at t')   (self-necessity of present & past)
∴ 3. inevitable at t' ($\sim$C at t) v inevitable at t' (R at t') (addition)
∴ 4. inevitable at t' ($\sim$C at t v R at t')
∴ 5. inevitable at t' (C at t → R at t')

And in any case surely what the determinist is wanting to claim about 'C at t → R at t'' is that it is necessarily true timelessly. For these reasons I use the ordinary timeless operator at this stage, though temporal determinants will be introduced later.

(iii) I have used the convention regarding t, t', t'' ... and t_0, t_1, t_2 ... that the latter denote definite times and the former any times, such that the interval t_0-t_1 is equal to the interval $t-t'$. Thus in (1), (2), (3), which state universal laws, the superscripts are used, and in ($3'$), (4), (5), which are about a particular case, the subscripts are used.

13   I leave the validity of this step to intuition at this stage.

14   Aristotle admitted this kind of necessity: 'there is no contingency in what has now already happened' (*Rhetoric* III 17, 1418a3–5); 'what is necessarily is, when it is, and what is not necessarily is not, when it is not' (*de Interpretatione* 19a23–5); 'it is not possible that what has come to pass not come to pass' (*Nicomachean Ethics* VI 2, 1139b7–9); 'there is no potency of having been, but only of being or going to be' (*de Caelo* I 12, 283b13ff.). The passage from *de Interpretatione* and that from *de Caelo* contradict each other on the question whether necessity of past fact belongs also to present fact. I think that this contradiction is to be resolved in favour of including the present in the necessary by observing that Aristotle must here have in mind a rather loose sense of 'present' in which it includes the short-term future, not the instantaneous present that we typically have in mind. Something of this sort must be true, for just a few lines earlier Aristotle writes 'For in its later state it (sc. an object which used not to exist but now does) will possess the capacity of not existing, but not of not existing at the time when it exists – since then it exists in actuality' (283b8–10). And this clearly implies that an object which exists at time t_n does not at t_n have the capacity of not existing at t_n: the thesis of the necessity of present fact.

15   Though Rescher and Urquhart on p. 5 of their book *Temporal Logic*, New York, Springer-Verlag, 1971, attribute such a system to Aristotle (without textual reference); and Wiggins, op. cit., uses some such apparatus, though he uses it as I have done, without formulating the whole system of which it would be a part.

16   *Analytica priora* I 9, 30a15–25. Alexander of Aphrodisias in his

commentary on this passage (124.8) records the dissent of Theophrastus and Eudemus from Aristotle's view.

17 Perhaps there lurk problems of existential import here, if one takes the view that A-propositions on the square of opposition lack such import but that singular ones have it. We can avoid such problems by making the unproblematic stipulation that there are some ps and some qs.

18 Jan Łukasiewicz, *Aristotle's Syllogistic,* 2nd ed., Oxford, Oxford University Press, 1957, pp. 183–4. ('A' in his notation denotes the universal affirmative, A-proposition, of the square of opposition.)

19 'Aristotle's theory of modal syllogisms and its interpretation' in M. Bunge (ed.), *The Critical Approach: Essays in Honor of Karl Popper,* New York, Free Press, 1963.

20 Storrs McCall, *Aristotle's Modal Syllogisms,* Amsterdam, North Holland Publishing Company, 1963, chapter 1.

21 Rescher, op. cit., section ix.

22 G. H. von Wright, 'A new system of modal logic' in his *Logical Studies,* London, Routledge & Kegan Paul, 1957, especially p. 112ff. The system is inspired by Aristotle's distinction between necessity 'haplos' and 'ex hypostaseos'. Patzig (*Aristotle's Theory of the Syllogism,* tr. J. Barnes, Dordrecht, Reidel, 1968) argues that this distinction is a mistaken one, that 'q is necessary on condition p' is a mistake for 'necessarily (p → q)'. In section (v) I shall introduce an apparatus which has among its incidental benefits that it will resolve this quarrel between Patzig and Aristotle. See n. 27, below.

23 Actually, von Wright offers three ways of reading M(p/q): (a) 'p is possible on conditions q', (b) 'p is possible relative to q', and (c) 'p is possible, given q'. Of these the third seems to me ambiguous between a conditional and a causal relation; but I find no evidence that von Wright had this point in mind or that the ambiguity was deliberate.

24 An axiom would be N(p:q) → q. It seems too that we should want to admit N(p:q) → N(p/q). These two, together with the inference [N(p/q) & q] → N(p:q) which we have already allowed, give us that [N(p/q) & q] ⟷ N(p:q), and we might as well allow that equivalence to provide the definition of N(p:q). What will *not* be permitted, and here the English is not reflected, is the inference N(p:q) → Np.

25 von Wright op. cit., pp. 121ff. The reference is to *de Interpretatione* 19a23–4 and 25–6. But it seems to me that Aristotle's point here cannot be made properly without the apparatus of temporal logic.

26 A law of this form (but without a distinction parallel to mine between □ and L) occurs in Łukasiewicz's Ł-modal system, expounded in the *Journal of Computing Systems,* vol. I (1953). It is, however, a many-valued logic.

27 This is the necessity which Aristotle denotes by 'anagke', see Patzig,

op. cit. and see n. 23 above. We can resolve the dispute between Patzig and Aristotle in the following way. The necessity which belongs to a formally valid syllogism is □; the necessity which belongs to the conclusion of such a syllogism when its premises are true is L. Aristotle is certainly wrong if he holds that □ belongs to both, and Patzig claims he does hold this; but Patzig strays needlessly far from the way in which people do in fact argue when he insists that there is no sort of necessity belonging to the conclusion as such.

28 Since it is a thesis of any modal system that $(p \rightarrow p)$, we have that $\Box(p \rightarrow p)$, and hence by law (6) that $(p \rightarrow Lp)$. This, together with law (2) gives that $(p \leftrightarrow Lp)$.

29 Since $(P \rightarrow D) \rightarrow L(P \rightarrow D)$ (self-necessity), we also have this inference schema validated:

(i) $p \rightarrow q$
(ii) p
∴ (iii) Lq

To raise the question whether this is too strong to be endured is to raise the question about the equivalence of $(p \rightarrow q)$ and $L(p \rightarrow q)$ – the old question about the difference between entailment and strict entailment. In this system there is no difference between $(p \rightarrow q)$ and $L(p \rightarrow q)$, though there is a difference between either of them and $\Box (p \rightarrow q)$.

The fact that (i), (ii), and ∴ (iii), above, is valid in this system has the convenience that we can get the necessary conclusion even if we are given the premise stating the law of nature, (i), merely as a regularity report: $p \rightarrow q$.

Disappointingly, this system will not really solve the problem about Barbara LXL; it will allow Barbara LXL to be valid only at the expense of allowing Barbara XXL to be valid as well. Barbara XXL will be valid by law (6); *a fortiori* Barbara LXL is valid. And the conclusion of either Barbara XXL or LXL can have its modality converted from *de dicto* to *de re* by law (5). To accommodate Łukasiewicz's striking tale about wires we must have Barbara LXL valid, but XXL invalid.

II ON THE DISPUTE BETWEEN COMPATIBILISTS AND INCOMPATIBILISTS

1 Professor Anscombe in her inaugural lecture says that compatibilism is 'gobbledygook'; and Davidson has a short-tempered moment in the first paragraph of his paper 'Freedom to act', writing, 'I know of none (sc. incompatibilist arguments) that is more than superficially plausible. Hobbes, Locke, Hume, Moore, Schlick, Ayer, Stevenson, and a host of others have done what can be done, or ought ever to have been needed, to remove the confusions that can make determinism seem to oppose freedom.'

Anscombe's lecture is *Causality and Determination,* Cambridge University Press, 1971; Davidson's paper is in *Essays on Freedom of Action,* Ted Honderich (ed.), London, Routledge & Kegan Paul, 1973.

2 A. J. P. Kenny, *Will, Freedom and Power,* Oxford, Blackwell, 1976, pp. 155-6.

3 P. Nowell-Smith, 'Ifs and cans', *Theoria,* 1960.

4 Ascensions, assumptions, levitations and wonders generally will be an embarrassment to this thesis if it is held that in them laws of nature are violated.

5 The same sort of asymmetry arises with knowledge. 'He does not know that p' is not the negation of 'He knows that p'. For the latter claims some things about his cognitive state and presupposes that p is true; but the former may deny either the claim about the cognitive state or the presupposition.

Of course if one takes the crude position that the presuppositions of a remark figure among its truth-conditions and that its truth-conditions amount to its meaning, to the claim it makes, then 'I cannot ϕ' is indeed the logical negation of 'I can ϕ'. 'I can ϕ' would lay a claim (a) to the logical or physical possibility of ϕing, and (b) to my having the skill of ϕing; 'I cannot ϕ' would disclaim either (a) or (b) or both. The negation of (a & b) is indeed ($\sim$a v $\sim$b). But it would still be true that where 'I cannot ϕ' means that ϕing is logically or physically impossible it follows that I do not ϕ.

There are of course some cases where it is not certain whether 'I cannot ϕ' claims physical impossibility or disclaims skill. 'I cannot hold a 100lb weight above my head for three minutes' is such a borderline case. The very interesting question that is posed by such cases is the question whether the uncertainty as to the meaning of the claim is merely epistemic.

6 It is important that this 'is' is that of contingent identity or material equivalence. Where the second premise of the inference states an implication only, whether contingent or not, the results are rather different. Thus, where ϕing is hitting the bull, and ψing is hitting the dartboard:

I can ϕ	I cannot ϕ
$\phi \rightarrow \psi$	$\phi \rightarrow \psi$
$\therefore$ I can ψ is valid	$\therefore$ I cannot ψ is invalid
I can ψ	I cannot ψ
$\phi \rightarrow \psi$	$\phi \rightarrow \psi$
$\therefore$ I can ϕ is invalid	$\therefore$ I cannot ϕ is valid

7 This example is adapted from Kenny, op. cit., who adapted it from Duns Scotus.

8 This particular argument was advanced to show the incompatibility between the ability to *do* otherwise and physiological determinism. The adjustments required to make it show the incompatibility between the ability to *decide* otherwise and neurophysiological determinism are straightforward adjustments.

9 Kenny, op. cit., pp. 132ff.

10 Or he could argue that its meaning is so exhausted but that 'opportunity' here is so construed that natural determinism removes it.

11 It is sometimes claimed that the 'or' of natural language is exclusive. But were that so in this case I should have to regard what I had said as false, and it is unlikely that I should want to say that.

12 Since she also advised him to seek Greek allies for his expedition.

III THE CORRELATION THESIS

1 E.g. D. H. Hubel & T. N. Wiesel 'Receptive fields of single neurones in the cat's striate cortex', *Journal of Physiology*, vol. 148, 1959.

2 W. Ritchie Russell, *Brain, Memory, Learning*, Oxford University Press, 1959.

3 Reported by Moyra Williams, *Brain Damage and the Mind*, Harmondsworth, Penguin, 1970, pp. 84ff.

4 A. R. Luria, *The Working Brain*, tr. Basil Haigh, Harmondsworth, Penguin, 1973.

5 'Could amount to sufficient truth-conditions for their asciption' is a deliberate form of words. I expect that many mental terms have for their truth-conditions a disjunction of sets of sufficient conditions, some of which sets include conscious states and others of which do not. After all we call other people adamant on the basis solely of their behaviour; we call ourselves adamant largely on the basis of certain conscious states. No one however would take seriously the idea that 'adamant' is ambiguous here. There seem to me to be few mental predicates which have, among the sets of sufficient truth conditions whose disjunction constitutes their meaning, even one set whose members are all conscious states.

6 Luria, op. cit., *passim*.

7 Donald Davidson, 'Mental events', in L. Foster and J. W. Swanson (eds), *Experience and Theory*, University of Massachussets Press, 1970.

8 H. R. Harré, *The Principles of Scientific Thinking*, London, Macmillan, 1970, pp. 209ff.

9 Heretofore we have spoken of mental *events*; nothing is concealed in the introduction here of the term 'state'; I perceive no relevant difference between the two and will use whichever of the two makes the more manageable and euphonious English.

10 This argument also commits us to the view that the number of different possible mental state kinds, though enormous, is finite – unless of course we allow that mental events can be of infinite length, like sentences.

11 In the remainder of this chapter, when I speak of brain states and mind states, I shall in general mean kinds of brain state and kinds of mind state; in cases where I intend individual states rather than kinds, this fact will be obvious.

12 Bernard Katz, *Nerve, Muscle, and Synapse,* New York, McGraw-Hill, 1963, p. 110.

13 Using Harré's principle of taxonomic priority of the mental in a rather transmuted form.

14 I believe that the project would be even more hopeless than this, for *contra* Wittgenstein and many others I believe that the range and variety of our mental life far exceeds the capacity of our language to describe that life. I shall not actually make this strong claim, however, for I do not here want to embroil myself in its defence, and anyway the weaker claim will serve my purpose as well.

15 There could be no species inferior to one which individuates as far as it is possible to individuate without recourse to techniques of numerical individuation: mental states (rather like Aquinas' angels) have almost no accidents – on this account only temporal ones.

16 D. M. Armstrong, *A Materialist Theory of the Mind,* London, Routledge & Kegan Paul, 1968, pp. 71–2.

17 The times do not coincide for presumably the replica is made to undergo what the man's CNS underwent *after* the CNS underwent it.

18 This hypothesis would, I think, inevitably lead one to deny the Lockean metaphysics of primary and secondary qualities. When asked, for example, about one's first yellow-experience why it was yellow, one could not say 'because yellow light is of frequency 5.2 $\times 10^{14}$ cps, and that frequency selectively excites some rods and cones of the retina, which transmit excitation to the area N4 (say) of the occipital lobe, and excitation of that area just does produce yellow-experience'. That answer would do only if one adhered to the Universal Preordination hypothesis; rather one would have to say, 'it was yellow because what was in the world just was yellow'. And that answer denies the view that secondary qualities – or at any rate *yellow* – do not exist in the world. At least, I cannot think of any other way in which the adherent of the Progressive Accidental Matching hypothesis could answer the question.

19 Thus far, it doesn't confirm a *Universal* Preordination hypothesis; but neither does it disconfirm it.

20 I am grateful to Dr Marc Colonnier for advancing this objection.

21 Professor Anscombe makes this point in her inaugural lecture, *Causality and Determination,* Cambridge University Press, 1976, p. 26.

22 Though it would well account for very extreme cases of akrasia.

23 There is lurking hereabouts the problem how to specify the mental language on the one hand and the neural on the other. I have no solution to offer for this problem as it has traditionally been undertaken. But I should like to make two remarks about it. The first is that this distinction is one for which we have very strong intuitions and for which there seem to be few problem cases. When

our intuitions are as strong as this the setting up of decision procedures for these languages is not urgent, and is undertaken out of tidiness of mind only. The second remark is that in any case we have an easy apparatus for solving the problem for the artificial language in use on the basic correlation tables we have envisioned: the very artificiality of the language makes it possible to set up a system for distinguishing them. We could stipulate, for example, that the numbers designating *infimae species* of mentals be preceded by the letter T, and that the numbers designating extended total neural states be preceded by the letter N.

24 Donald Davidson, 'The individuation of events', in N. Rescher (ed.), *Essays in Honor of Carl G. Hampel* Dordrecht, Reidel, 1969, p. 231ff.

25 The causal criterion includes, or entails, the criterion of spatio-temporal coincidence: it seems to be true that any two events which have all their causes and all their effects in common must occur in the same place at the same time. This is an incarnation of Hume's principle of the spatio-temporal contiguity of causes and their immediate effects. This principle has new life now that 'causation at a distance' is in disrepute, with the wave-theory of gravitation.

26 Sheet lightning is said to ripen corn.

27 Another way to deal with the counter-example of the phantom limb and the pain in the toe is to note that not all events in fact occur where they seem to occur: ventriloquism provides a striking example of this. It could be argued that pains in the toe misleadingly seem to occur in the toe, and that pains in the phantom wrist misleadingly seem to occur in the air. (As far as we know there is no phenomenon of phantom trunk or phantom head.)

28 Jerome Shaffer, 'Could mental states be brain processes?' *Journal of Philosophy,* LVIII, 1961.

29 E. J. Lemmon has argued similarly in his comments on Davidson's paper in N. Rescher (ed.), *The Logic of Decision and Action,* University of Pittsburgh Press, 1966.

30 If one chooses to adopt a regularity view of the causal relationship, then there is nothing impossible about interactionism. But on the other hand the apparent impossibility of interactionism is good evidence that the regularity theory does not match our intuitions of what causality is.

31 In case (b) the neurophysiological determinist's thesis is not one of universal determinism, for his argument will not work for first occurrences of mental event kinds. But is freedom of decision something we want to maintain for babies and young children?

32 Hereinafter when I speak of the Correlation thesis, I shall, unless I say otherwise, intend a version of that thesis which is somehow strong enough to allow the neurophysiological determinist to advance at least a fairly universal determinism.

IV THREE PROBLEMS FOR THE LIBERTARIAN

1 This point matches my explanation in chapter I, part 1, section (iii) that I was happy to accept incomplete determinism, that is, to argue for 'incomplete' freedom.
2 Or rather, by some physicists.
3 In *The Neurophysiological Basis of Mind,* Oxford University Press, 1953, p. 278ff., and an improved account in *Facing Reality,* London, English University Press, 1970, p. 125.
4 *Facing Reality,* p. 12.
5 *The Neurophysiological Basis of Mind,* p. 273ff.
6 As far as I can make out it was first advanced by Cicero against the Epicurean doctrine of *clinamen,* or swerve.
7 J. R. Lucas, *The Freedom of the Will,* Oxford University Press, 1970.
8 This point was made by Kenny in the Gifford Lectures of 1971-2, H. C. Longuet-Higgins, A. J. P. Kenny, J. R. Lucas, C. H. Waddington, *The Nature of Mind,* Edinburgh University Press, 1972, p. 75.
9 Jonathan Glover offered this succinct refutation in his book *Responsibility,* London, Routledge & Kegan Paul, 1979, p. 31.
10 The most recent statement of the argument was in his contribution to a symposium the proceedings of which were edited by J. Eccles under the title *Brain and Conscious Experience,* Berlin, Springer-Verlag, 1966.

V HEGEMONY

1 Thus far this is meant to be ambiguous as between the case in which an agent raises his arm and the case in which someone else raises it for him.
2 The description of this event in terms of the movements of muscles and tendons must not be too precise or the desired ambiguity (n. 1 above) will be lost: 'his arm went up' will differ for the two cases: (a) he raises his arm, and (b) his arm was raised by someone or something else.
3 That is, not: I raise my arm by holding it with my other arm and lifting it, as one does, for example, when the arm has gone to sleep.
4 Strictly, of course, it is not a case of descriptions explaining descriptions but of the occurrence of the event under one description explaining its occurrence under the other. This is too cumbrous, however, and I shall use the shorter, less accurate, formulation.
5 Though I admit that this game is sufficiently odd that it would not be surprising to find that players of it had odd motives.
6 Much psychoanalytic explanation is notoriously of this weak type, and one may well doubt whether it is improvable. The standard plots of primal dramas seem to admit a variety of quite opposite

denouements. A cold mother can as well cause coldness - by imitation, as gushiness - by reaction, in her child. A weak father may cause effeminacy in a son because he fails to counterbalance the mother's influence, or the son may become aggressively masculine in compensation. And so forth.

7 This is as far as our present purposes require us to take this theory of hegemony and strength of explanation. One can however envision the next step in refinement, and it is metaphysically instructive. Suppose a given event is rendered as necessary under two quite different descriptions. An example might be a case of loss of temper. We could explain it by saying that a person of such a disposition had to explode when threatened in such a persistent way. We could also explain it neurophysiologically and say that such and such a neural state triggered the rage-reaction inevitably. Here I believe that we should find the latter explanation stronger, for the reason, I suggest, that its vocabulary is more fine-grained. We have a strong proclivity to believe that all vagueness is epistemic, that the truest descriptions are the highly detailed ones, that reality always bears analysis.

8 Not only is the decision, at the mental level, not necessitated, it is not even (typically) rendered uniquely by teleological explanation.

9 In one version of this the reasonings must occur before the decision; in another all that is required is that the agent be prepared to give reasons. The former would preclude many decisions from being free ones - whimsical ones, for example; the latter is surely too weak - one can be prepared to give reasons for many acts that were not one's own free acts.

10 D. Wiggins, 'Towards a reasonable libertarianism', in Ted Honderich (ed.), *Essays on Freedom of Action,* London, Routledge & Kegan Paul, 1973.

VI THE RANDOM AND THE FREE

1 *Nicomachean Ethics,* 1110a1ff., 1110a15ff., and 1110b1ff. For simplicity I here disregard the complication about ignorance.

2 *Physics,* 241b28, 243a11.

3 *de Caelo,* 268a28.

4 *Physics,* 254b15ff. A self-mover can suffer unnatural motion, of course, but not *qua* self-mover; i.e. the unnatural motion would derive from something else.

5 *Physics,* 255a5ff.

6 He says so himself at 254b33.

7 *Physics,* Book II, Chapters 5 and 6.

8 *Physics,* 254b30, 255a15.

9 *Physics,* 255a13.

10 *Eudemian Ethics,* 1224b25ff.

11 It may seem unfair to confront Aristotle with such a modern

example, but there is evidence that he knew things of the same general kind. In *de Motu animalium,* 701b2ff. he seems to be discussing toy carriages that work on similar principles. I do not mean to say that Aristotle would cheerfully allow that such engines are self-movers whose movements are voluntary; indeed he explicitly restricts self-movement to the animate, as we have seen. My point is rather that for all he says of the *structure* of self-movement, its restriction to the animate seems arbitrary.

12 Richard Taylor, *Action and Purpose,* Englewood Cliffs, N.J., Prentice-Hall, 1966, pp. 15ff. I do not in the pages following pretend to be discussing Taylor's whole position in any careful way. Rather I pick up a suggestion of his and develop it in my own way, and against the particular challenge of neurophysiological analysis of the agent.

13 This case is discussed in detail in chapter VII, section (i), below.

14 Professor Anscombe may be said to have been discussing it in her inaugural lecture, *Causality and Determination,* Cambridge University Press, 1971.

15 Another example of this which seems especially interesting has to do with atomic theory and the differing densities of substances. We regard differing densities of substances as a feature which is to be explained by their composition out of differing proportions of atoms and void. Atomic theory lies so deep with us that when we read in a non-atomist like Aristotle that densities are primitive, and when we really think about this idea, it seems impossible. Of course it is not *impossible* that densities be primitive, but to think of them as primitive requires shedding our atomic theory of matter, and that theory is so fundamental in us that it is extraordinarily difficult to shed. All important revolutions in thought require this labour of shedding.

16 Thomas Reid repudiates the notion of 'passive power' in no uncertain terms. In treating of Locke he writes, 'I do not remember to have met with the phrase *passive power* in any other good author. Mr. Locke seems to have been unlucky in inventing it; and it deserves not to be retained in our language.' *Essays on the Active Powers of the Human Mind,* Baruch Brody (ed.), Cambridge, Mass., MIT Press, 1969, p. 23.

17 This position is substantially that of Reid, ibid., pp. 4–12.

18 Not all members of a species have the same active powers. All human beings have the power to decide, I suppose; most have the power to raise their arms; a few also have the power to wiggle their ears. Studies of the effects of 'biofeedback' have shown that it is possible to acquire new powers of this sort, like the power to lower the temperature of one's hand. It is not entirely clear whether this should be regarded as the acquisition of a new power or as the discovery of a power one already had. I suppose one should (following Aristotle, *de Anima,* II, 5) say that there are two stages of power or potentiality, the power proper and the

power to acquire the power proper.

19 Of course, it would have to pass through the agent in a certain way: an electric current through an agent which made his fingers curl, for example, would not count as activity. But we can ignore these niceties here. See chapter VII, section (i), below.

20 Leibniz, *Textes inédits*, G. Grua (ed.), Paris, 1948, vol. II, p. 512. The translation of the definition is by Martha Kneale, and is taken from her paper 'Leibniz and Spinoza on activity' in Harry G. Frankfurt (ed.), *Leibniz*, New York, Doubleday, 1972.

21 *Leibniz: Philosophical Writings*, tr. Mary Morris, London, Dent, 1934, pp. 211–12.

22 There is a good discussion of this by John Beloff in 'The identity hypothesis: a critique' in J. R. Smythies (ed.), *Brain and Mind*, London, Routledge & Kegan Paul, 1965, pp. 47ff.

23 Some of the extraordinary difficulties of this concept are deployed in his *Being and Nothingness*, tr. Hazel Barnes, New York, Washington Square Press, 1966, pp. 125ff.

24 *de Generatione animalium*, 726b22.

VII PARALIPOMENA

1 This experiment is described, by, for example, Leonard A. Stevens, *Explorers of the Brain*, New York, Knopf, 1971, p. 261.

2 A middle case which would repay study is addiction.

3 I do not mean to use this term in any precise legal sense.

4 Aristotle, *Nicomachean Ethics*, 1110a27.

5 C. A. Campbell, *Selfhood and Godhood*, London, Allen & Unwin, 1957, pp. 149ff.

BIBLIOGRAPHY

Alexander Aphrodisiensis, *In Analytica priora* in the series *Commentaria in Aristotelem Graeca*, Berlin, Prussian Academy 1882–1909.
Anscombe, G. E. M., 'Causality and determination', Cambridge University Press, 1971.
Aristotle, *The Works of Aristotle Translated into English*, vols I–XII, ed. W. D. Ross, Oxford University Press, 1908.
Armstrong, D. M., *A Materialist Theory of the Mind*, London, Routledge & Kegan Paul, 1968.
Ashby, W. Ross, *Design for a Brain*, London, Chapman & Hall, 1965.
Ayer, A. J., 'Man as a subject for science', London, Athlone Press, 1964.
Ayers, M. R., *The Refutation of Determinism*, London, Methuen, 1968.
Beloff, John, 'The identity hypothesis, a critique' in J. R. Smythies (ed.), *Brain and Mind*, London, Routledge & Kegan Paul, 1965.
Bergson, Henri, *Time and Free Will*, tr. F. L. Pogson, London, Allen & Unwin, 1950.
Berofsky, Bernard (ed.), *Free Will and Determinism*, New York, Harper & Row, 1966.
Binkley, Robert, Bronaugh, Richard and Marras, Ausonio (eds), *Agent, Action and Reason*, Oxford, Basil Blackwell, 1971.
Borst, C. V. (ed.), *The Mind/Brain Identity Theory*, London, Macmillan, 1970.
Broad, C. D., 'Determinism, indeterminism and libertarianism', in *Ethics and the History of Philosophy*, London, Routledge & Kegan Paul, 1952.
Bunge, Mario (ed.), *The Critical Approach: Essays in Honor of Karl Popper*, New York, Free Press, 1963.
Campbell, C. A., *Selfhood and Godhood*, London, Allen & Unwin, 1957.
Campbell, C. A., *In Defence of Free Will*, London, Allen & Unwin, 1967.
Chomsky, Noam, *Language and Mind*, New York, Harcourt, Brace & World, 1968.

Bibliography

Craik, Kenneth, *The Nature of Explanation,* Cambridge University Press, 1967.

Danto, A. C., 'Representational properties and mind-body identity', in *Review of Metaphysics,* vol. 26, 1973.

Danto, A. C., *Analytical Philosophy of Action,* Cambridge University Press, 1973.

Davidson, Donald, 'The individuation of events', in N. Rescher (ed.), *Essays in Honor of Carl G. Hempel,* Dordrecht, Reidel, 1969.

Davidson, Donald, 'Mental events', in L. Foster & J. W. Swanson (eds), *Experience and Theory,* Amherst, Mass., University of Massachusetts Press, 1970.

Davidson, Donald, 'Freedom to act', in Ted Honderich (ed.), *Essays on Freedom of Action,* London, Routledge & Kegan Paul, 1973.

Davis, William H., *The Freewill Question,* The Hague, Martinus Nijhoff, 1971.

Dray, W. H., *Laws and Explanation in History,* Oxford University Press, 1970.

Eccles, J. C., *The Neurophysiological Basis of Mind,* Oxford University Press, 1953.

Eccles, J. C., *The Physiology of Synapses,* Berlin, Springer-Verlag, 1964.

Eccles, J. C., *Facing Reality,* London, English University Press, 1970.

Eccles, J. C., *The Understanding of the Brain,* New York, McGraw-Hill, 1972.

Eccles, J. C. (ed.), *Brain and Conscious Experience,* Berlin, Springer-Verlag, 1966.

Evans, C. R. and Robertson, A. D. J. (eds), *Key Papers in Brain Physiology and Psychology,* London, Butterworths, 1966.

Ferré, Frederick, 'Self determinism', *American Philosophical Quarterly,* vol. X, July 1973.

Fodor, Jerry A., *Psychological Explanation,* New York, Random House, 1968.

Foster, L. and Swanson, J. W., *Experience and Theory,* Amherst, Mass., University of Massachusetts Press, 1970.

Frankfurt, Harry G. (ed.), *Leibniz,* New York, Doubleday, 1972.

Franklin, R. L., *Freewill and Determinism,* London, Routledge & Kegan Paul, 1968.

Gifford Lectures 1971–2, *The Nature of Mind,* A. J. P. Kenny, H. C. Longuet-Higgins, J. R. Lucas, C. H. Waddington, Edinburgh University Press, 1972.

Glover, Jonathan, *Responsibility,* London, Routledge & Kegan Paul, 1970.

Goldman, Alvin I., *A Theory of Human Action,* Englewood Cliffs, N.J., Prentice-Hall, 1970.

Harré, H. R., *The Principles of Scientific Thinking,* London, Macmillan, 1970.

Harré, H. R. and Secord, P. F., *The Explanation of Social Behaviour,* Oxford, Basil Blackwell, 1972.

Bibliography

Harré, H. R. and Madden, E. H., *Causal Powers*, Oxford, Basil Blackwell, 1975.

Hart, H. L. A., *Punishment and Responsibility*, Oxford University Press, 1973.

Hart, H. L. A. and Honoré, A. M., *Causation in the Law*, Oxford University Press, 1959.

Hintikka, Jakko, *Time and Necessity*, Oxford University Press, 1973.

Honderich, Ted (ed.), *Essays on Freedom of Action*, London, Routledge & Kegan Paul, 1973.

Honoré, A. M., see Hart, H. L. A.

Hook, Sidney (ed.), *Determinism and Freedom*, New York, Collier, 1961.

Hubel, D. H. and Wiesel, T. N., 'Receptive fields of single neurones in the cat's striate cortex', *Journal of Physiology* (London), vol. 148, 1959.

Huby, Pamela, 'The first discovery of the freewill problem', *Philosophy*, vol. 42, 1967.

Katz, Bernard, *Nerve, Muscle and Synapse*, New York, McGraw-Hill, 1966.

Kenny, A. J. P., *Action, Emotion and Will*, London, Routledge & Kegan Paul, 1963.

Kenny, A. J. P., *Will, Freedom and Power*, Oxford, Basil Blackwell, 1976.

Kenny, A. J. P., *Freewill and Responsibility*, London, Routledge & Kegan Paul, 1978.

Kenny, A. J. P., see also Gifford Lectures.

Laslett, Peter (ed.), *The Physical Basis of Mind*, Oxford, Basil Blackwell, 1968.

Leibniz, *Textes inédits*, ed. G. Grua, Paris, 1948.

Leibniz, *Philosophical Writings*, tr. Mary Morris, London, Dent, 1934.

Lemmon, E. J., 'Comments on Davidson's "The Logical form of action sentences" ' in N. Rescher (ed.), *The Logic of Decision and Action*, University of Pittsburgh Press, 1967.

Lucas, J. R., *The Freedom of the Will*, Oxford University Press, 1970.

Lucas, J. R., see also Gifford Lectures.

Łukasiewicz, Jan, *Aristotle's Syllogistic*, 2nd edn, Oxford University Press, 1957.

Łukasiewicz, Jan, 'A system of modal logic', in *Journal of Computing Systems*, vol. 1, 1953; reprinted in *Jan Łukasiewicz Selected Works*, ed. L. Borkowski, Amsterdam, North Holland Publishing Company, 1970.

Luria, A. R., *The Working Brain*, tr. Basil Haigh, Harmondsworth, Penguin, 1973.

McCall, Storrs, *Aristotle's Modal Syllogisms*, Amsterdam, North-Holland Publishing Company, 1963.

MacKay, D. M., 'From mechanism to mind' in J. R. Smythies (ed.), *Brain and Mind*, London, Routledge & Kegan Paul, 1965.

Madden, E. H., see Harré, H. R.

Marras, Ausonio, see Binkley, Robert.

Marshall, W. A., *The Development of the Brain*, Edinburgh, Oliver & Boyd, 1968.

Morgenbesser, Sidney and Walsh, James (eds), *Free Will*, Englewood Cliffs, N.J., Prentice-Hall, 1962.

Munn, A. M., *Free Will and Determinism*, London, MacGibbon & Kee, 1960.

Nowell-Smith, P., 'Ifs and cans', *Theoria*, 1960.

Patzig, G., *Aristotle's Theory of the Syllogism*, tr. J. Barnes, Dordrecht, Reidel, 1968.

Popper, Karl, 'Indeterminacy in quantum mechanics and in classical physics', *British Journal for the Philosophy of Science*, vol. 1, 1950.

Popper, Karl, 'Indeterminism is not enough', *Encounter*, April 1973.

Prior, A. N., 'Limited indeterminism', *Review of Metaphysics*, vol. 16, 1962 (reprinted in his *Papers on Time and Tense*, Oxford University Press, 1968).

Reid, Thomas, *Essays on the Active Powers of the Human Mind*, ed. Baruch Brody, Cambridge, Mass., MIT Press, 1969.

Rescher, Nicholas, 'Aristotle's theory of modal syllogisms and its interpretation', in M. Bunge (ed.), *The Critical Approach: Essays in Honor of Karl Popper*, New York, Free Press, 1963.

Rescher, Nicholas (ed.), *The Logic of Decision and Action*, University of Pittsburgh Press, 1966.

Rescher, Nicholas (ed.), *Essays in Honor of Carl G. Hempel*, Dordrecht, Reidel, 1969.

Rescher, Nicholas and Urquhart, Alasdair, *Temporal Logic*, New York, Springer-Verlag, 1971.

Robertson, A. D. J., see Evans, C. R.

Royal Institute of Philosophy Lectures, *The Human Agent*, London, Macmillan, 1968.

Russell, W. Ritchie, *Brain, Memory, Learning*, Oxford University Press, 1959.

Sartre, J.-P., *Being and Nothingness*, tr. Hazel Barnes, New York, Washington Square Press, 1966.

Secord, P. F., see Harré, H. R.

Shaffer, Jerome, 'Could mental states be brain processes?' *Journal of Philosophy*, 1961, reprinted in C. V. Borst (ed.), *The Mind/Brain Identity Theory*, London, Macmillan, 1970.

Smythies, J. R. (ed.), *Brain and Mind*, London, Routledge & Kegan Paul, 1965.

Stevens, Leonard A., *Explorers of the Brain*, New York, Knopf, 1971.

Strawson, P. F., 'Freedom and resentment', *Proceedings of the British Academy*, vol. xlviii, 1962 (reprinted in his *Freedom and Resentment and Other Essays*, London, Methuen, 1974).

Swanson, J. W., see Foster, L.

Taylor, Charles, *The Explanation of Behaviour*, London, Routledge & Kegan Paul, 1964.

Taylor, Richard, *Action and Purpose*, Englewood Cliffs, N.J., Prentice-Hall, 1966.

Urquhart, Alasdair, see Rescher, Nicholas.

Vivian, Frederick, *Human Freedom and Responsibility*, London, Chatto & Windus, 1964.

Walsh, James, see Morgenbesser, Sidney.

Wiesel, T. N., see Hubel, D. H.

Wiggins, David, 'Towards a reasonable libertarianism' in Ted Honderich (ed.), *Essays on Freedom of Action*, London, Routledge & Kegan Paul, 1973.

Wilkes, K. V., *Physicalism*, London, Routledge & Kegan Paul, 1978.

Williams, Moyra, *Brain Damage and the Mind*, Harmondsworth, Penguin, 1970.

Wooldridge, Dean E., *The Machinery of the Brain*, New York, McGraw-Hill, 1963.

Wright, G. H. von, *Logical Studies*, London, Routledge & Kegan Paul, 1957.

Young, J. Z., *A Model of the Brain*, Oxford University Press, 1964.

INDEX